James Henry Bennet

On the Treatment of Pulmonary Consumption

By Hygiene, Climate, and Medicine

James Henry Bennet

On the Treatment of Pulmonary Consumption
By Hygiene, Climate, and Medicine

ISBN/EAN: 9783337155766

Printed in Europe, USA, Canada, Australia, Japan

Cover: Foto ©berggeist007 / pixelio.de

More available books at **www.hansebooks.com**

ON THE TREATMENT

OF

PULMONARY CONSUMPTION

BY

HYGIENE, CLIMATE, AND MEDICINE.

WITH

AN APPENDIX

ON THE SANITARIA OF THE UNITED STATES, SWITZERLAND,
AND THE BALEARIC ISLANDS.

BY

JAMES HENRY BENNET, M.D.

MEMBER OF THE ROYAL COLLEGE OF PHYSICIANS, LONDON;
LATE OBSTETRIC PHYSICIAN TO THE ROYAL FREE HOSPITAL; DOCTOR OF MEDICINE OF THE
UNIVERSITY OF PARIS;
FORMERLY RESIDENT PHYSICIAN TO THE PARIS HOSPITALS (EX-INTERNE DES
HÔPITAUX DE PARIS), ETC. ETC.

"Medio tutissimus ibis."—*Ovid.*

THIRD EDITION.

LONDON:
J. & A. CHURCHILL, NEW BURLINGTON STREET.
1878.

PREFACE

THE SECOND EDITION.

THE first edition of this essay was published in 1866, and was purely clinical. Although deeply interested in the discussion then going on respecting the intimate nature and origin of Tubercle and of Pulmonary Consumption, I thought it best to confine myself entirely to the clinical aspect of the disease. I merely described, therefore, the hygienic or sthenic treatment, which gives results infinitely more favourable than those obtainable by the antiphlogistic treatment generally adopted in my younger days.

Such reticence, however, is no longer desirable or possible. Two doctrines are in presence, both based on minute histological research, and both exercising, all over the world, a profound influence on the treatment of Pulmonary Phthisis. The one—that of Virchow and his followers—all but denies the tubercular origin of chronic Pulmonary Consumption—of the disease clinically recognised as Pulmonary Phthisis; in their eyes it is either a new growth or a mere form of chronic pneumonia. The other—represented by Professor Bennett of Edinburgh—defends and supports the tubercular origin of chronic

b

Pulmonary Phthisis, in an exudation of the blood
previously impoverished, also on histological grounds,
entirely repudiating the data on which Virchow and
his school found their opinions.

The singular feature of this controversy is, that the
immense body of practising physicians, those who
have to act as umpires between the two histological
schools, are not competent even to give an opinion on
the subject of minute microscopical morbid anatomy
discussed.   They are followers, not leaders, in microscopical research, and have to see with the eyes of
faith what they are told to see by the teachers in
whom they trust.   They are wanting in the knowledge or sharpness of eyesight which would enable
them to detect who is right and who is wrong, when
such men as Bennett and Virchow, who have spent
their lives in microscopical research, thoroughly dis-
agree.

Practical clinical physicians have, however, it
appears to me, a *locus standi*, a basis on which to
rest their decision, and that basis is the clinical observation of the disease in question, the one that I
took in 1866.   It is this basis that I have again
taken as my guide, and which leads me to lean rather
to the doctrine of my namesake than to that of the
celebrated German pathologist.   I would draw attention to the fact that not only is Professor Bennett
a distinguished Histologist, but he is also equally
distinguished as a Pathologist.   For more than a
quarter of a century he has taught clinical medicine

at the bedside to a numerous class, and he is universally known as a clear-sighted observer. It is this
combination of histological knowledge with that
acquired by the lengthened observation of disease
that, in my mind, gives peculiar and exceptional
value to his opinions.

In leaning to the view respecting the origin of
Pulmonary Phthisis advocated by the Edinburgh
Professor, I am gratified to find that I agree entirely
with Dr. Austin Flint of New York, one of the most
eminent and experienced of our American brethren
(see p. 115).

I believe also that I am warranted in stating that
most physicians of matured age and of enlarged
experience who have given themselves the trouble to
study this controversy, lean to the same side—and
are indisposed to substitute the term " chronic pneumonia" for that of " chronic Phthisis," and that on
the ground of long-continued clinical observation.
On the other hand, most of the followers of the new
"inflammation doctrines" are young, enthusiastic
physicians, full of knowledge, but deficient in
experience.

As we advance in life our experience naturally
increases. Thus the mental horizon expands, and
much more is seen, in Medicine as in everything
else, than in the earlier part of life. It would be
sad indeed if, as we lose power, as our sight fails, we
did not reap some little compensation.

As a traveller who has seen many climes, who has

ascended many Alpine ranges, it often strikes me
that man's mental vision as he advances in life may
be compared to the ocular vision of the traveller who
ascends a mountain.   He starts, at early dawn, full
of life, strength, and enthusiasm, but at first, at the
base of the mountain, he sees but little, only what is
near and around him.   Thus he makes many mistakes, falls into many quagmires.   As the day progresses he rises higher and higher, and simultaneously
his vision expands.   He no longer merely sees the
objects just around him ; he grasps those far off—the
road he has traversed, that which he has still to
encounter, the ravines and dangers passed, and some
—and some only—of those that have still to be met.
But the higher he gets, the more he sees, the more
his strength flags.

So it is in Medicine.   As our experience of disease
increases, when we have followed its developments in
several generations of human beings, we learn to
take more comprehensive views, to attach less importance to local morbid manifestations, more importance to general laws, to hereditary, constitutional,
social, hygienic conditions.   It is, after all, to this
comprehensive clinical experience of disease that we
must appeal in such discussions as the one which is
reviewed in this essay.

THE FERNS, WEYBRIDGE, SURREY,
    Oct. 1871.

PREFACE

TO

THE THIRD EDITION.

SEVEN years have elapsed since the Second Edition of this work appeared, and they have been to me years of deep thought on the subjects of which it treats, as also of active clinical work. The experience of each successive year, however, has only confirmed the convictions therein recorded. I have nothing to alter, merely to endeavour to give greater clearness and force to their expression.

In the year 1874 I reproduced the work in French, in Paris, and I am told that the views enunciated have tended to modify considerably the current of professional feeling, and the treatment of Phthisis, in France. If so, I have not laboured in vain. I have now added a sketch of the views entertained by the leading French pathologists respecting the intimate nature of Phthisis.

I have devoted a chapter to the consideration of mountain climates, and have given in an Appendix an account of recent travels in Switzerland, undertaken with a view to elucidate this question, an important one, which much occupies the thoughts of the profession.

I cannot conclude without paying a passing tribute to the memory of my old friend and companion in life, Professor Bennett of Edinburgh, whose name is so often mentioned in this work. It will, I believe, remain connected with the literature of Phthisis. He passed away two years ago, at the age of sixty-two, in the zenith of his intellectual power, regretted by the entire profession, at home and abroad. He has left me, his invalid friend, whose days he thought were numbered twenty years ago, and other practical observers, to continue the battle he so vigorously initiated. Be it so.

THE FERNS, WEYBRIDGE, SURREY,
Oct. 1878.

CONTENTS.

CHAPTER VI.

CHAPTER VII.

CHAPTER VIII.

———

ON THE

TREATMENT OF PULMONARY CONSUMPTION

BY

HYGIENE, CLIMATE, AND MEDICINE,

IN ITS CONNEXION WITH MODERN DOCTRINES.

CHAPTER I.

PRELIMINARY REMARKS.

THE NATURE AND CAUSES OF PULMONARY PHTHISIS.

So much has been written during the last thirty years on Pulmonary Consumption by men of the highest order of intellect, that it requires a certain amount of moral courage to enter the arena. No one, indeed, would be justified in so doing unless he conscientiously thought that he had information to impart calculated to be of use to his fellow practitioners, and to humanity at large. It is because I believe that such is really my case, as regards the treatment of pulmonary phthisis, that I have written the present clinical essay.

I have little that is absolutely new to bring forward—at least, that is new to those who keep pace with the progress of Medicine ; but I believe that I have important testimony to give in favour of modern science, as my experience powerfully corroborates

B

modern views with reference to the treatment of
pulmonary consumption. One who, like myself, has
been forty-four years in the profession, belongs both
to the past and to the present, and is able to speak
from personal experience of the views and opinions of
former days as well as of those of the present time.
This fact may give some weight to the judgment of
an author, who certainly does not appear in the character which, according to the Latin poet, charac-
terizes the later periods of life—that of a " laudator
temporis acti."

I believe, also, that I have had rather peculiar
and exceptional opportunities for forming an opinion
respecting the value of the treatment of pulmonary
consumption pursued thirty or forty years ago as
compared with that now adopted by the leading
authorities at home and abroad. My education was
carried out partly in England, partly in Paris.
During four years (1839-43) I was resident medical
officer to several Paris hospitals, and gave clinical
lectures on auscultation (not then as generally
studied as it is now) and on diseases of the heart
and lungs, to several hundred young English and
American medical men. Thus I became thoroughly
imbued with the knowledge of the day, and have ever
since remained much interested in thoracic pathology.

After practising many years in London, as an obstetric physician, I became myself affected with consumption, and in 1859, had to abandon everything
in order to go and die, as I thought, on the shores of

the Mediterranean. Relieved from the fatigue, the harass, and the cares of our arduous profession, I managed, by the application of modern science, to save my own life, and since then I have helped to save the lives of many similarly affected who have followed me in my winter health-exile to the south.

The great fact to which I have to testify is, that pulmonary consumption is often a curable disease—indeed, in its early stages, often a very curable disease—under sthenic or restorative treatment.

In making this assertion, I have merely to confirm by matured knowledge, a paper written in 1840, entitled " On the Curability of Consumption," which I recently found among my manuscripts. At that date I had just passed a year as one of the resident medical officers of the Salpétrière, a large asylum hospital in Paris, more especially devoted to aged and infirm women. There I had found in the dead room, in the lungs of many women who had died in advanced life from other diseases, large cretaceous deposits and puckered cartilaginous cicatrices, which proved emphatically, undeniably, that they had been consumptive at some antecedent period of their life, but had got well, spontaneously no doubt, dying at last of other disease. Indeed, in those days, the rational treatment of Phthisis was so little understood by the generality of practitioners, that I believe a sufferer had a better chance of recovery if the disease was not discovered than if it was. The low diet, the confinement, the opiates and fever medi-

cines, the leeches and blisters, which constituted the
usual therapeutics of such cases, were certainly but
little calculated to arrest a disease the essence of
which is organic debility. The inhabitants of the
Salpêtrière are mostly aged women belonging to the
lower classes, and the pathological conditions observed were undeniable evidence of their having
recovered from an advanced stage of consumption, at
some period or other of their life. Possibly, feeling
ill and weak, they had taken refuge with their relations or friends in the country, and had recovered
under the influence of improved hygienic conditions,
or of a natural limitation of the disease.

In the year 1840, when, from my own personal
observation, I became aware of these pathological
facts, that Pulmonary Phthisis is spontaneously curable, it was considered by most of the physicians with
whom I had come in contact to be all but inevitably
fatal, especially when advanced to its second stage—
that of softening. Since then, the same pathological
evidences of the spontaneous cure of Phthisis have
been found by all who have had special opportunity
of making post-mortem observations in the aged.
Among these observers I may more especially mention my namesake, Professor Bennett of Edinburgh.*

---

* Pulmonary Consumption, by John Hughes Bennett, Professor of
the Institutes of Medicine, Edinburgh University. First edition, 1853;
2nd edition, 1859, p. 26.

Clinical Lectures on the Principles and Practice of Physic. Edinburgh, 1868; fifth edition, p. 179.

Article, Phthisis Pulmonalis, in Reynolds's System of Medicine.
Vol. iii. p. 537. 1871.

His luminous work on Pulmonary Consumption, (1853), has much contributed to improve our knowledge of this disease. He was the first to thoroughly establish its treatment on the rational ground of hygiene. Indeed, Professor Bennett has done more than any other British physician to base the treatment of Phthisis on a true appreciation of the nature of the disease, and to prove that it is capable of arrest and even of cure, by a restorative treatment, if recognised before the lungs are so disorganized as to be unable to discharge their physiological functions.

This, I believe, is now the opinion of most practitioners, at home and abroad, whose attention has been specially directed to the subject. Among those who early laboured to demonstrate this fact I may name my friend Dr. Richard Quain, one of the most enlightened physicians of the day. In 1852 he published a series of very interesting and valuable cases, illustrating, in the most unimpeachable manner, the arrest and cure of Pulmonary Phthisis under sthenic treatment.* There are many, however, who still look upon Pulmonary Phthisis as incurable, or who are wedded to old and erroneous methods of treatment

I would remark, incidentally, that the modern change in the treatment of Pulmonary Phthisis is entirely the result of doctrinal progress—of the gradual establishment of pathological views sounder and more rational than those that reigned at the

---

* See *Lancet*, 1852, May 22nd, June 12. Report of Cases from the Out-patient Department of the Hospital for Consumption, Brompton.

beginning of this century. I do not believe in the change of the type of disease as a cause of the change of therapeutic doctrines. In my experience disease is exactly what it was when I was a youth, forty-four years ago, allowing, of course, for epidemic influences, whilst my own ideas have undoubtedly changed. The contrary opinion has always appeared to me a mere figment, brought forward by men advancing in years, who had rather believe that disease has changed than that they themselves can have held false doctrines, and can have treated their patients on erroneous data, in youth and middle age. The probability is that disease is pretty much now what it was a thousand years ago, or in the days of the patriarchs, modified of course by hygienic and social conditions.

Pulmonary Consumption was thought, by most pathologists, in the early part of this century, to be merely a disease of the lungs, having an intimate connexion with inflammation. The doctrines of Broussais were then the order of the day. Nearly every disease of the human economy was considered to be inflammatory, or the sequela of inflammation, and a depleting, lowering, "antiphlogistic" treatment was adopted.

Under the influence of microscopical research, and of more minute and searching investigations into morbid anatomy, these views, doctrinal and therapeutic, have been altogether modified. A lowered vitality, a lowered vital power, are generally con-

sidered, and justly so, to be at the root of most
morbid changes, and a sthenic or life-supporting
instead of a depleting treatment is recommended and
adopted in the treatment of disease generally.

The recent histological views of Virchow and his
followers have, however, a tendency to revive the
Broussaisian explanation of Phthisis, to refer it to
the class of inflammatory diseases. As such views
must exercise an influence over treatment, I cannot
avoid alluding to them, or asking, What is the
intimate nature of Phthisis Pulmonalis—what is
meant by the term Pulmonary Consumption?

The answer to this question is both easy and
difficult. Easy if we merely consider the clinical
features of the disease so called ; difficult if we try to
reach its intimate nature. The general consensus of
the medical profession embodies under these terms
the forms of lung disease in which morbid deposits,
localized or general, take place in the intimate structures of the lungs, extend from one region to another,
soften gradually, destroy and excavate the lungs, and,
if unchecked, end in the death of the patient. So
explained and limited, the disease stands out clinically clear and well defined. If, however, we try to
ascertain the exact nature of these morbid deposits
or formations, obscurity and doubt begin. To some
they are, generally, tubercular deposits or exudations,
to others they are, generally, inflammatory or scrofulous products, only exceptionally tubercular. To
make these views clear, however, I must refer to the

text of the authorities. This is the more necessary, as although the intimate and structural nature of tubercle on the one hand, and of the products of inflammation on the other, has been most minutely investigated by hosts of competent observers, there still appears to be room for doubt and discussion.

The views of the more eminent histologists appear to group themselves round two schools—viz., that of Professor Bennett and that of Virchow.

According to Professor Bennett (Lectures, p. 179), " Tubercle may be regarded as an exudation, possessing deficient vitality, sometimes grey, but more frequently of a yellowish colour, varying in size, form, and consistence, essentially composed of molecules and irregularly-formed nuclei. . . . This exudation" (p. 181) " has little tendency to pass into cell forms. The original albuminous molecular matter melts into nuclei, which constitute the tubercle corpuscles, and are developed no further. . . . . I regard tubercle, therefore, as an exudation which may be poured out into all vascular textures, in the same manner, and by the same mechanism, as occurs in inflammation ; only from deficiency of vital power it is incapable of undergoing the same transformations, and exhibits low and abortive attempts at organization, and more frequently, as a result, disintegration and ulceration. For the same reason we observe that whenever an undoubted inflammation becomes chronic, with weakness, the symptoms and general phenomena become identical with those of tuberculosis. Hence there is little difference between a chronic pneumonia of the apex of a lung and a phthisis ; the one indeed passing into the other. . . . . When we endeavour to discover the ORIGIN of the weakness producing this effect on the exudation, we must ascribe it to imperfect nutrition."

Dr. Reginald Southey,* in a series of Lectures exhaustive of the histology of Tubercle, as described by German histologists, gives the following account of Virchow's doctrines :—

" Tubercle" (p. 12), "according to Virchow, is a new growth, and is classed by him among the Lymph tumours—those, namely, which are constructed after the pattern of lymph glands, and which stand most closely in relation to connective tissue formations. The single tuberculum or the tuberculous tumour is not capable of identification from any one element entering into its composition ; but its origin, its development, and its minute structure together confer a particular stamp upon it as a whole, which renders it capable of distinct recognition.

" The Tubercle formation is cell-structured from the moment of its first appearance; it proceeds always out of connective tissue, or from some tissue closely allied to this, such as false membrane, fat, or the medullary tissue of bone. It exists in two forms —the one the cellular, the other the fibrous form ;

---

* The Nature and Affinities of Tubercle : being the Goulstonian Lectures for the year 1867. London : Longmans. 1867.

but they have such features in common as imply unmistakable oneness. The fibrous form is only a
slight structural modification of the simple cellular—
a modification impressed upon it by the external
conditions of growth.

" The origin and mode of development of the simple
cellular form is best of all to be studied from the
tuberculous growth, as this is found upon serous
membranes, or upon the mucous membrane of the
larynx. . . . . On a serous membrane the young
growth is smaller than the smallest millet-seed ; it
has a granular look, and contains soft, imperfectly-
developed cells, which are very easily broken down,
and free nuclei. Its elements, although differently
grouped, are identical with those that constitute a
normal lymph gland.

" The isolated Tubercle forms the tiniest tumour
that occurs on the human body, but it is rarely, if
ever, single. These growths are almost always
multiple ; they are found in nest-like groups close
together, and multitudes of nodules thus originally
and individually distinct can combine together and
form a conglomerate tumour.

" Conglomerate Tubercle increases in size by surface accretion : new nodules grow up in the tissue
immediately round about the old ones ; and as this
accretion can take place from all sides around a centre
in the parenchyma of solid organs, the final shape
attained by the mass is round or roundish."

According to Virchow the formations, exudations,

deposits, infiltrations, which take place in the lung in a chronic form, principally in the apices, are not, generally speaking, tuberculous—that is, formed of tubercles, as above described. They are, generally, either bronchial lymphatic gland tumours of a scro-fulous nature assuming the form of caseous or cretaceous masses ; or they are the result of scrofulous (grey) hepatization of the lungs.

" The catarrhal process" (p. 84) " extends into the ultimate air-cells, the appearances finally attained being scarcely distinguishable from those which attend acute pneumonia ; catarrhal mucus cells and fibro-plastic ovoid nuclei are mixed together and block up the alveoli ; the inter-vesicular spaces are trebled in thickness, and come to present a dense, and more or less fibrillated connective tissue ; . . . . . in the thoracic tract," (p. 78) " the implication of the lungs, a gland tissue of peculiar aim, and therefore of very special structure in scrofulous chronic inflammation, gave Louis first, then Laennec, and after him the French schools, their primary conception of tuber-cular disease ; and as Virchow and others maintain, a false idea of the constitution and essential nature of Tubercle.

" If, as in the neck, the scrofulous process had ex-pended itself less upon its primary seat and more upon the bronchial glands (which although large do not become proportionately enlarged) its pathology probably would have been more correctly under-stood. The peculiar, round, millet-seed-like cast of the ultimate alveoli and finer bronchi, formed of inspissated catarrhal mucus, which shot out like an adventitious product upon gentle pressure, gave colour to the deception by its likeness to actual Tubercle of the pleura and serous surfaces; and the not infrequent occurrence of the two diseases together in the same person (for they in no way exclude each other), led to this mucus cast of an air-sac, for I can call it nothing else, being thought to be the *fons et origo malorum*, the typical and primary form of Tubercle. . . . . It was an adventitious product—it was produced by inflammation; true enough it was the result of a local inflammatory process. It was deposited from the blood, and coagulated in the part where it was found; it softened into a cheesy mass, which hollowed from its centre. As Carswell showed, it was the almost entire plugging of small bronchi which gave rise to this idea. Vomicæ were produced by the colliquating influence of the softened tubercles, by the death from pressure and cutting off of nutritive supplies of the interposed parts, and extended by the inflammation of surrounded tissues thus aroused."

Virchow, according to Dr. Southey, appears to reserve the term tuberculosis, all but exclusively, to those cases in which undeniable tubercle is formed simultaneously throughout the economy, or throughout an entire organ, as in the meninges of the brain in tubercular meningitis, or in the lungs in acute Phthisis. Such cases are acute in their manifestation and development, are accompanied by severe constitutional disturbance, and are all but necessarily fatal in their termination.

From the above it will be seen that the schools represented by Professors Bennett and Virchow differ in their account of the histological origin of Tubercle in the following manner. Professor Bennett thinks that tubercle is exuded from the blood, depraved by defective nutrition. Professor Virchow deems that tubercle is formed by a new growth from the connective tissue, the result of irritation or *Reiz*, as he terms it, originating in this connective tissue. The difference between these two views is most material, and clinically most important, when we come to the consideration of the chronic, sparse or conglomerated deposits, infiltrations, of the lung tissue, which occur in Pulmonary Consumption.

Virchow considers them to be new growths from the fibrous tissues, resulting from irritation or inflammation, whereas Professor Bennett thinks them generally to be coagulated exudations from the blood, which has previously been rendered poor by defective nutrition.

In France, German pathology has, of late years, gained ground, especially in all questions to the solution of which histology is appealed, and the prevailing doctrines respecting the intimate nature of Phthisis incline to those of Virchow and Niemeyer. The erudite work of Messrs. Herard and Cornil entitled, "De la Phthisie Pulmonaire, étude anatomique, pathologique, et clinique," Paris, 1870, may be considered to embody modern French views. The authors seem to occupy a medium position between the two histological schools, that of Virchow, and that of the late Professor Bennett of Edinburgh. They accept granular or miliary tuberculisation as the point of departure, perhaps constant, of chronic pulmonary Phthisis, but look upon the chronic deposits which are found in the lungs as the result of catarrhal broncho-pneumonia, of lobular pneumonia, which they describe under the name of *caseous pneumonia*.

At page 526, under the heading of "Comparative Study of the Different Forms of Pulmonary Phthisis," I find several paragraphs which embody the doctrine of these authors ; I translate textually.

The symptomatic study to which we have just devoted such lengthened and minute development has furnished, we hope, the demonstration of two important facts which it was the object of these clinical researches to establish—viz., that in all the varieties of pulmonary Phthisis identical lesions are found (granulations, pneumonia), that the form of the malady is essentially determined by the extent of these lesions, by their combination in different proportions, and at the same time by the rapidity of their evolution.

The first of these propositions has been abundantly proved by the details of pathological anatomy into which we have entered.

As to the *granulation* more especially, we have seen that it does not constitute a lesion outside of tuberculisation (*granulie*), that it does not either belong to a particular form of tuberculisation (*acute Phthisis*), but that it was found in all the admitted varieties, as well in the chronic form as in the sub-acute, acute, galloping forms, &c. We repeat that if the external appearance of these granulations, their volume, their histological composition, appear to differ, it is only because they are examined at different epochs of their evolution. But at the onset they always present themselves under the form of grey nodosities, semi-transparent, the dimensions of which, at first very small, microscopic, increase insensibly, at the same time that they become opaque in the centre, and then quite yellow. If, in chronic tuberculisation, this last aspect of the granulations is more frequently met with, it is that these granulations are older, and have had time to pass by all the phases of their development, and to arrive at the last stage of their metamorphoses, the cellulo-adipose retrogression. In the most acute forms, in those most rapidly fatal, such as general granular Phthisis, the granulations are the same, they would have undergone the same transformations had not death arrested their evolution. This is so true, that when the disease is less rapidly fatal, in the rare cases in which one lung only is attacked with general granular Phthisis, or when the lungs are not simultaneously invaded, as in some observations of M. Colin, we see side by side with grey and half transparent granulations, other granulations opaque, yellow, caseous, absolutely as in chronic tuberculisation. In the latter, also, we often meet, in the midst of old products, recent development of granulations with the external characteristics of young granulations, in the inferior lobe of the lung, for instance, or again on the visceral or thoracic pleura.

As to the pneumonia, it is the same, whatever the form of tuberculisation presented to the observer. It is always that catarrhal lobal or lobular pneumonia, the intra-alveolar products of which are re-absorbed with extreme difficulty, and pass after a variable period into the caseous state, so as to represent those yellow, more or less voluminous masses, sometimes extending to a great part of the parenchyma (the tubercular infiltration of authors), or those islands disseminated in the lung, which more especially respond to what is erroneously designated *crude tubercle.* This passage from catarrhal pneumonia to caseous pneumonia takes place with more or less rapidity. In the galloping form it is exceptionally rapid ; in general, it demands a certain lapse of time, and that is why the caseous state is so common in chronic tuberculisation, so rare, on the contrary, in the pneumonic form of general granular Phthisis. Death ensues in this case with such rapidity, from the extent of the morbid conditions, that inflammation of the lungs has not the time to pass beyond the period of congestion, and of red, lobular hepatisation, otherwise it would also arrive at the caseous phase. What shows that the morbid conditions are the same is, that sometimes we observe in this form one or more caseous nodules at the summit, or even sometimes small excavations, which indicate a morbid condition of older date, and of the same nature as the pneumonias of the chronic form. What still further proves that these lesions are identical, is that if, from any cause, death is delayed, the pneumonia has the time to pass through the last stages of their development, and the caseous state appears."

I am certainly incompetent myself to form an opinion on so minute and recondite a question of histological morbid anatomy. "Non nobis tantas componere lites." It is, however, of extreme importance *clinically*, that the profession should know that all these histological conceptions imply the same disease, Pulmonary Consumption. For the want of such an understanding most palpable and lamentable clinical errors are being daily committed, much to the detriment of the poor sufferers.

Whatever the intimate nature of these chronic deposits in the lungs, whether really tubercular or merely inflammatory, one fact is certain—that they constitute the disease described by Laennec and Louis as Phthisis Pulmonalis. To call them merely caseous pneumonia, or inflammatory consolidations of lung tissue, or lymphoid tumours, as is now very generally done in Germany, France, and England, or

C

" lung troubles," a term often used by our American brethren, is practically a deception and a snare; unless the old-fashioned idea of " Phthisis," of a disease distinctly and naturally tending to progress and death, is connected therewith.

Unfortunately, when the above terms are used they are often not so connected mentally, either by the young physician, imbued with recent histological lore, or by the patient deluded by terms, the true meaning of which he does not understand. Every winter I see at Mentone a number of decidedly consumptive patients from different parts of the Continent, or of America, often physicians, who say that they have nothing much amiss—only a slight inflammatory consolidation of the lung, a slight caseous pneumonia of the apex, a scrofulous abscess in the lungs, or some slight " lung trouble." Yet these patients are often far advanced in pulmonary consumption, with all its local ravages and constitutional symptoms.

I often find them in the third stage of Phthisis, with large cavities and with only a few months to live. Generally speaking, they have not the slightest idea that death has laid his hand on them, that their ambitious social career is at an end, that the ship is in danger, that it is only by throwing overboard the cargo, by making the most desperate and continued sacrifices that they can hope to prolong or to save life.

If, therefore, such terms as: localized pneumonia, caseous pneumonia, inflammatory consolidation of a limited area of the lung, scrofulous abscess or deposit, lung trouble, are to be retained and used by the profession, it is our duty to explain to the patients that they are merely modern phrases for Pulmonary Consumption—for Phthisis. Otherwise it is utterly impossible to induce them to take that care of themselves, to make those sacrifices, which are indispensable for their recovery.

As the chronic malady of the lungs in which the tubercular exudations or formations, the inflammatory products, the scrofulous caseous tumours above described, constitute the disease universally recognised and described as Pulmonary Consumption or Phthisis, we cannot do better than retain for it the latter name. By so acting we avoid assuming, as decided, points still obscure, or at least still debated. At the same time we imply, as of old, that we are dealing with a very serious and fatal disease, with one which, unchecked, has an undeniable tendency to progress and to destroy life, once it has reached a certain stage of its development; that we have to treat a dire malady by which a considerable portion of the human race passes into eternity. There can, however, be no objection to applying the term tuberculosis to general acute tuberculization, were it thus applied by general consent. In England the general mortality from Phthisis is twelve per cent.—in other words, one death in eight is from Phthisis. According to Dr. Farr's reports, which coincide with our knowledge on the subject, this general mortality is very

much greater in towns than in the country. Any
sedentary occupation at once raises the rate of
Phthisisical mortality, whereas agricultural out-door
pursuits lessen it.

Whatever opinions be now entertained as to the
intimate nature of the morbid conditions of lung
which constitute Pulmonary Phthisis, I think that
there can be but one opinion as to their secondary
character. Indeed, it is generally admitted, as
forcibly demonstrated by Professor Bennett in 1853,
that the lung disease, however severe, chronic or
acute, is merely an epiphenomenon, the secondary
result of constitutional disease, of a morbid general
diathesis which precedes, occasions, and rules the
lung malady. Given the lowered state of vitality,
given the diathesis it engenders, these morbid tubercular or other formations or deposits may take place,
as a result, in the lungs, and in all but any other
organ or region of the body.

Thus, in the investigation of the nature and causes
of Pulmonary Consumption, and of tubercular disease
in general, we must go a step further than can be
reached by the most minute anatomical and histological researches. Clinical observation shows that the
manifestation of these forms of lung disease, chronic
and acute, must be looked upon as the evidence
and result of a serious, perhaps final, diminution of
vital or nervous energy. In other words, they may
be considered the evidence of incipient decay of the
organization from defective vital or nervous power.

Indeed, Pulmonary Phthisis and general tuberculization are simply a "mode of dying." Unless the vitality of the individual can be roused, the morbid condition will surely progress, and life will be extinguished, sooner or later according to the state of the constitution of the patient, and to the type of the disease.

Forty years ago the doctrine of Broussais reigned paramount in therapeutics, and the treatment was consequently antiphlogistic, debilitating. The disease was considered to be one in which the inflammatory element predominated, and to be all but inevitably fatal. The sufferers were treated under the influence of these ideas, and nearly all died. Since then (1840) therapeutics have changed. The more enlightened physicians of our own time have adopted a rational vitalism. They have understood and accepted that in all diseases it is an error to weaken the patient in order to kill the disease; that we must, on the contrary, not only spare his vital force, but endeavour to increase it. In a word, the antiphlogistic doctrines of the first part of this century have been laid aside.

As I have already stated, it has been asserted by many distinguished physicians, at home and abroad, that these changes of medical and therapeutical doctrines are the result of a change of type in disease. According to them disease is no longer what it was, and the sufferers themselves are no longer what they were forty years ago. Formerly diseases were inflammatory, sthenic, now they are anæmic asthenic.

I do not accept this view of pathology, past and present. For me disease is exactly what it was nearly half a century ago, when I commenced the study of medicine. I see no difference whatever in it or its symptoms; it is I who have changed, not pathology. I have sedulously endeavoured to revive my recollections of early practice, vividly impressed on the memory, I have reperused my note-books, studied published cases, and still can perceive no difference.

As to Pulmonary Phthisis, as I have already stated, most of the patients, if not all, whom I saw treated, or treated myself, in the early period of my medical career, according to the antiphlogistic doctrines of the day, which I then held, died. A considerable number, on the other hand, of those whom I have seen and treated during the last twenty years, according to the views of rational vitalism developed in this book, have regained health, or at least greatly prolonged their existence.

The German histological doctrines of our own times, having a direct and inevitable tendency to bring therapeutics back to the ideas which formerly allowed sufferers to die who were curable, are dangerous doctrines in a clinical point of view, especially for young physicians. However well informed a young physician may be, if he thinks that he has only to treat a catarrhal pneumonia, a lobular pneumonia, a chronic caseous pneumonia, he is very far from looking upon the disease from which his patient is suffering, as a probable verdict of death. And yet it is only by taking this view of the case that he can hope to arrest its progress, still less to cure it.

The very essence of life is the organic vitality, variable in different species, variable in different individuals, with which each organism, vegetable or animal, merges into being and develops itself. It is owing to inherent organic vitality that the medium duration of life in the Oak, the Ash, the Fir, is different, as it is also different in the Whale, the Elephant, the Horse, the Dog, and in Man himself. The medium duration of life in each species is reached in the organisms that are created under favourable conditions, with unimpaired organic vitality, and that pursue their existence under conditions favourable to life. On the other hand, this medium duration is not reached by those individuals that are created under unfavourable conditions, with defective vitality, or in whom originally sound vitality is modified, diminished, destroyed by the unfavourable conditions in which their existence is carried on.

In such considerations, in my opinion, must we seek for the real explanation of Pulmonary Consumption and of the tuberculization in general ; as also for a key to the types under which the disease presents itself and to the results of treatment. They include, of course, hereditary predisposition.

Viewed in this light, so far from Pulmonary Consumption being a dire inexplicable pestilence, striking

indiscriminately the young and the old, it becomes
one of the provisions by which Providence has
secured the integrity of the human race. If those
who are, from birth or otherwise, sickly or weak, in
whom vitality is defective originally, or secondarily
and accidentally, could propagate their kind so that
their progeny could live, the human race would soon
degenerate and become a race of pigmies, of sickly
dwarfs, and eventually die out. Pulmonary Phthisis
is, in reality, one of the diseases by which Providence
eliminates those who are weak, imperfect, and consequently unfit to perpetuate the race in its integrity.
Individually it may be very hard to be thus eliminated for the good of the human race ; but if we
rise above individuals and grasp the interests and
well-being of the entire human family, it will be seen
that these diseases are, in truth, a bountiful dis-
pensation of Providence. They may be compared
to hurricanes in tropical climates, which purify
the earth and contribute to make it habitable,
although often at the expense of great individual
suffering.

A man or woman who is old, who has inherited
disease, who labours under disease, or has become
weakened by disease, by privation, by cares—parents,
in a word, in whom organic vitality is weakened—
cannot give strong or even medium vitality to their
progeny. No one can give to others what he does
not possess himself. It is the same with plants. The
seeds of a young, vigorous plant produce healthy,

.vigorous plants ; whereas the seeds of old, weak, sickly plants produce a like progeny.

The human, like the vegetable, progeny, may at first be· fair to look at—may appear sound and vigorous ; but this state does not last. It is a mere deception ; for the inherent, the inherited vitality is defective. Such beings are like bad watches made with bad works ; they may look well, and go well for a time ; but they soon wear out, go badly, and, if patched up again, get out of order, and finally stop. The good watch, on the contrary, made with good works, will go a hundred years ; or, if it accidentally gets out of order, it will go as well as ever when once it has been set to rights. Thus is explained death by Pulmonary Consumption at the age of fifteen, twenty, thirty, of young people born with defective vitality, even when brought up and living in favourable conditions for life, and apparently healthy and vigorous. They have thus early exhausted the amount of vitality which they received from their parents. They have used to the last shred their constitutional powers, and decay, commencing in the shape of Pulmonary Phthisis, closes their earthly career—unless their vitality can be roused by rational hygienic treatment.

Those who are fortunately born with a fair amount of organic power, their progenitors being young and healthy, may damage it as they advance in life. Unfavourable hygienic conditions, accidents, cares, the thousand incidents of the struggle of life, may

impair their originally good constitution, and diminish.
or even crush their vitality. When such is the case,
death may come in a hundred ways, through the
attacks of a hundred diseases ; but one of the most
common modes of decay, especially in towns, is Pulmonary Consumption. In town life the supply of
atmospheric air is generally defective in quality, and
this, I am convinced, is one of the most efficient immediate causes of phthisis.

In both cases—whether Pulmonary Consumption
attacks persons deficient in inherited organic vitality,
or whether it attacks those in whom originally sound
vitality has been accidentally diminished by the wear
and tear of life—it is in reality a secondary element.
It is not the real disease, but a symptom of it. The
real disease is exhausted or lowered vitality. Thus
when a tree in a forest or plantation is attacked by
insects, fungi, and parasites of all kinds, they are
only *apparently* the cause of its decay and death.
The young, vigorous, healthy, tree resists their attack
through its high vitality ; full of life, it fears no such
enemies. If they attack its less fortunate companion,
it is because the latter is already sickly, diseased.
The true remedy is not merely to scrape away the
moss and to kill the parasites, for others will come ;
but to remove all causes of ill-health and decay : in
a word, to rouse the vitality of the tree by trenching
and draining the soil, by putting good loam round
its roots, and by protecting it from all injurious
influences. Thus only can we hope to succeed in

arresting the decay. If we are successful, the tree will itself gradually shake off its enemies, and may even eventually be restored to pristine vigour and beauty.

Such, I believe, should be the treatment of Pulmonary Consumption. There is no panacea whatever for a disease which is merely a symptom of lowered vitality, of positive decay. But much may be done towards arresting the progress of this decay, and even towards effecting a cure, by the combined influence of hygiene, of climate, and of rational medical treatment. These are the three modes of treatment which I intend to discuss, and that in the order given—the order of their relative importance.

CHAPTER II.

HYGIENE.

BODILY HYGIENE : FOOD—STIMULANTS—RESPIRATION AND VENTILATION — THE SKIN — EXERCISE — MENTAL HYGIENE : THE PASSIONS.

IF, as I have stated in the preceding chapter, the exudations, deposits, or formations in the intimate tissues and structures of the lungs which constitute Phthisis, are the result of defective nutrition, consequent on defective vitality, inherited or acquired, the rules for treatment become self-evident—they must be found principally in the strict observance of the laws of physiology and hygiene. In most cases of this disease it will be discovered, on careful inquiry, that these rules have been grossly infringed.

The laws of hygiene may be considered to embody the conditions, bodily, social, and mental, which are the most favourable to the healthy development of the human economy, the most conducive to its well-being. These conditions have only been clearly elucidated by modern research, and are daily ignored and infringed by the immense majority of the human race -- with comparative impunity by the strong, the vigorously constituted, but not so by the weak, by those who are born with defective vitality, or are living in unhygienic conditions. In both cases existences, strong or weak, which might have reached or exceeded the ordinary term of human life under favourable conditions, are prematurely brought to a close, and very often by Pulmonary Phthisis.

Bodily hygiene includes, principally, good and abundant food, pure air, a clean skin, and exercise. Mental hygiene includes rational, not extreme, mental exercise, and the regulation of the passions.

Theoretically, the injunction to attend to bodily hygiene in a disease of debility seems so rational that it appears scarcely necessary to lay stress upon it ; but practically it is not so. A large proportion of the medical profession, instead of looking upon the progressive increase of morbid deposits in the lung, with their gradual softening, upon the hæmorrhages, and the bronchial and laryngeal affections which they occasion or which precede them, as mere local symptoms of a general diathesis, have their attention arrested by the local condition. They exaggerate its inflammatory nature, and dare not apply to their patients the ordinary rules of hygiene ; they dare not give wine and plenty of animal food ; they dare not give fresh, cool air day and night ; and they dare not keep the skin clean and cool by cold or tepid sponging. Yet this timidity is a fatal mistake, for these are the principal means by which nutrition is to be improved and restored to a normal condition, and consequently by which the disease is to be arrested and cured.

The food taken by consumptive patients should be of the most nourishing kind—meat, fish, fowl, eggs, milk, bread—well cooked, and abundant in quantity. Indeed, the quantity of food taken should principally be limited by their digestive powers. In my opinion, the principal value of medical treatment in phthisis is in the restoration of digestive tone when impaired or absent. If patients can be brought to eat, to digest, and to assimilate, they have a chance of recovery. If they cannot, their chance is indeed slight.

The medical attendant, however, must never forget an important fact often ignored, which I have developed in my work on "Nutrition,"* p. 150—viz., that there are, both in childhood and in the more advanced stages of life, two great types of digestive power, the quick and the slow. In many persons, in most indeed, the digestive process is rapid. Such individuals require food often—three or four times in the twenty-four hours, and in that period they can take and digest animal food two or even three times. If they have not these meals they often feel faint and ill. The other class digest more slowly, more laboriously. They can only take food, with advantage and comfort to themselves, twice, or at the most three times, in the twenty-four hours,

* Nutrition in Health and Disease: a Contribution to Hygiene and to Clinical Medicine. 8vo, p. 260, 3rd Cabinet Edition. 1877. Churchill.

and only one meal must be a meat meal. Such individuals become dyspeptic if they try to assume the
habits of those who require food more frequently.
The real remedy for their dyspepsia is not physic,
but the adoption of a dietary more suited to their
constitution. These peculiarities remain in disease,
and must be met, for the patient to do well; there
is no rule but the patient's own individual constitution. It is worthy of remark that the people who
thrive on two meals, get more out of their food, evidently, than those with quicker digestions.

A moderate amount of wine, as a tonic and gentle
stimulant to digestion, is undoubtedly beneficial—
say from four to eight ounces; that is, two or three
glasses of Claret, Burgundy, Hock, or two of Sherry,
taken with meals, and diluted with water; a better
plan than taking it alone; or a glass of bitter beer,
with or without water or soda-water, according to
constitution and to individual peculiarity. Of late
years, in America, alcohol, especially whisky, has
been much lauded as a remedy in Consumption. I
have seen a certain number of cases in which it had
been long taken, but I cannot say with benefit.
The daily ingestion of large quantities of nerve-stimu-
lating, carbon-producing spirit certainly does not
come under my idea of hygienic treatment. (See
"Nutrition," p. 100.) Carbonaceous food can be
given to any extent in a more simple form. When
alcohol is prescribed medicinally there is always the
risk of abuse. It is a double-edged sword.

In most constitutions, moreover, what may be termed the spirit limit—that is, the limit at which alcoholic beverages cease to do good, and are actually pernicious—is easily reached. An ounce, or an ounce and a half, two or three tablespoonfuls of actual spirit in the 24 hours, appears to be the usual limit. If an over-dose has been taken—an over-dose for the individual—digestion, instead of being benefited, is actually disturbed. That it is so is proved by interrupted or heavy sleep, a dry, clammy, blanched or furred tongue in the morning. On awakening the mouth in health ought always to be pure and sweet, and the tongue should be like that of a healthy child. There ought to be no sordes between the teeth, no desire to wash the mouth or to brush the teeth to be comfortable. When these symptoms exist, apart from the reaction of actual disease, the dietary of the previous day, and especially the amount of alcoholic beverage taken—beer, wine, or spirits—should be carefully investigated. It will often be found that some error is being committed, and this study of the individual is generally the best key to the amount and nature of the food and of the alcoholic fluids which ought to be taken.

In estimating these facts, the examination of the urine and of its deposits is of inestimable value, as I have explained at length in the work quoted, "Nutrition." The urine ought to be clear when passed, and to remain clear on cooling, whatever the temperature, be it warm or cold, when the individual is not under the influence of tangible acute or chronic disease. If, after the ingestion and digestion of food it is cloudy on passing, or if, being clear at first, it becomes cloudy and deposits salts on cooling, it is quite certain that the food has been imperfectly digested, that chylification has been defective, and the dietary and habits of life should be carefully analysed and studied. This rule even applies in a chronic disease such as Pulmonary Consumption. There is infinitely more to be done by the detection and rectification of individual diet errors than by any medicinal agencies. In plain common sense, what can a few grains of some drug or other do to counteract morbid digestive conditions resulting from and kept up by individual dietetic errors, daily committed, daily renewed ?

A simple but nourishing dietary may be insisted on, within the limits of reason, under all and every condition of the lung, with softening or without, with fever or without, with local inflammatory complications or without. We must try to struggle through unfavourable stages and complications without letting down nutrition.

When, however, serious complications exist in other organs, such as the liver, the kidneys, or the uterus, it is often all but impossible properly to nourish a consumptive patient. The tongue remains hopelessly coated, the appetite is completely in abeyance. As long as the complications exist this defective state of digestion and nutrition is irremediable. This is frequently the case when uterine complications exist.

In the year 1865 I published some papers in the *Lancet*, in which I stated that I not unfrequently found females whom I attend in the south for Consumption to be suffering also from uterine disease. Through the usual morbid reaction of the uterus on the stomach they have no appetite, are tormented with nausea, and can neither eat nor digest. They all invariably perish if I am not able in time to remove the uterine malady, and thus to restore the tone of the stomach, the natural desire for food, and the power of eating and digesting. Sometimes even when this has been accomplished, and they can eat and digest, it is too late. The disease has progressed too far ; the lung is all but gone, has become a mere shell, and the patient sinks a victim rather to uterine disease than to consumption, the apparent cause of death.

Scarcely a winter passes without my seeing several such cases—cases in which the Pulmonary Consumptive disease is complicated by extensive uterine lesions, and in which Phthisis is clearly secondary, the sequela and indirect product of the uterine malady. This serious complication is more especially met with in married women who have had several children, and in whom the uterine mischief is connected with past abortions or confinements. The prominent symptom in their case is the chest suffering, so they naturally apply to pure physicians, not practising or studying

midwifery or uterine diseases. Unfortunately many
really distinguished general physicians are unacquainted with the more obscure forms of uterine
suffering, and do not recognise them when met with.
Indeed some seem not to believe in the existence of
any other uterine disease than cancer and tumour,
or their attention is so concentrated on the chest that
other morbid conditions pass unperceived. Thus,
one of my great difficulties with such patients is the
doubt and incredulity occasionally shown by their
medical attendants at home as to the uterine complications which I discover. My special and exceptional
knowledge of these diseases, as a well-known and
successful author, as an old obstetric hospital physician, as a London consultant in female diseases of
thirty-five years' standing, all is ignored. I am thus,
not unfrequently, considered to be riding a hobby by
men who have never studied or attended to the
subjects in question. I could narrate various remarkable cases of this character that have passed
under my care, but will confine myself to the following one.

In November, 1869, I was consulted by a lady,
married, aged thirty, mother of several children, sent
to Mentone by her London medical attendants for
advanced Phthisis. Her case was indeed considered
so advanced, so bad, that the possibility of her going
at all was discussed at home. One of her physicians
wrote to me that he doubted her reaching her desti-
nation alive. On her arrival at Mentone she was so

exhausted that she had to be carried upstairs, and
to take to her bed, where I found her a week afterwards. She was thin, emaciated, feverish, sleepless,
had night sweats, no appetite, nausea. Indeed had
her case been purely phthisical she must have been
considered in the latter stage of the disease. I found
phthisical deposit and infiltration at the apex of the
left lung, posteriorly and anteriorly down to the level
of the second rib, with softening. There was no
cavity, and no evidence of lung disease elsewhere.
Clear as was the local evidence of consumptive
disease, the constitutional disturbance appeared to
me greater than was warranted by the extent and
stage of the actual mischief, and by the fatigue of
travelling, so I pursued my inquiry. I found that
she had had four or five children very rapidly, had
had a miscarriage three months before, was supposed
then to be pregnant owing to the absence of the
menses, and had abundant purulent leucorrhœa, with
a host of uterine symptoms. To me, and to all obste-
tricians acquainted with the present state of science,
the case was perfectly clear and self-evident. On
examination I found, as anticipated, extensive uterine
disease—a patulous hypertrophied cervix, a large
unhealthy ulcerated surface inside and outside the
os uteri, and other usual vaginal, uterine, and ovarian
complications.

Under the influence of proper surgical treatment
there was a marvellous rally. The digestive system
rallied as soon as the uterine lesions assumed a

healthy character, as often occurs in such cases. The nausea disappeared, the appetite revived, nutrition began to improve, and in a few weeks the patient was down stairs, sitting up at table, driving out— having resumed colour and a healthy hue of complexion. At the same time all the chest symptoms rapidly subsided—climate and cod-liver oil aiding; for now the oil, which the stomach previously could not bear, was taken, and borne with ease and great benefit. In this case the uterine lesions were steadily giving way under judicious and energetic treatment —as they often do when the result of confinements and abortions — when my hands were suddenly stayed. I received a communication from the principal London consultant, an eminent physician since deceased, stating that a consultation had been held on this lady's case, and that it had been decided to request me to suspend all uterine treatment, as they did not agree with me in my views. Of course I threw up the attendance, and this poor sufferer succumbed eventually to her sad disease. Perhaps she would have died anyhow; but perhaps, also, her death may have been the result of the ignorance and blindness of her home medical attendants. She lost a fair chance in the battle of life.

It seems at first sight as superfluous to state that in a disease of debility like Pulmonary Consumption patients should breathe pure air as that they should live on nourishing food; but it is not. Theoretically the value of pure air—of atmospheric food—is universally accepted by the medical profession : practically it is all but universally neglected. The physiology of respiration—a modern discovery—has yet to be applied, not only in every-day life, but even in the treatment of disease. Most medical men as well as their patients ignore the all-important fact that the demands of respiration are so great that one or two human beings soon use up and contaminate the air contained in a good-sized room. Such being undeniably the case, unless it be renewed artificially by an open window or by an open door or by both— in other words, unless the air in an inhabited room be constantly undergoing change—impure air is breathed, air calculated to produce disease even in the healthy, and to increase it in the sick. So universal is the neglect of this fundamental law of health, that the healthy persons who do not sleep in rooms with the windows, doors, and register stoves shut, and who do not thereby poison their blood all night with their own excreta, are as yet the exception.

In ill-health, and especially in diseases of the respiratory organs, the dictates of science and of common sense are still more grossly outraged. At a time when, perhaps, the principal food the economy can take is pure air, when the diseased lungs, partly inefficient, require the purest and best air-food that can be afforded them, the doors and windows are generally kept shut on pretence of chills, cold air, and draughts ; a due supply of respirable air being thus refused to the unfortunate patient.

In my younger days this fatal and cruel error was carried to an insane extent by many medical practitioners, as it still is in most parts of the Continent, and especially in Germany. The windows were often hermetically shut, and paper pasted over the chinks. The doors were made double, and one always shut before the other was opened. The healthy friends of the patient considered it a penance and a trial to have to remain in the polluted atmosphere "necessary" for the miserable sufferer, and often paid for their devotion by the loss of their own lives. On the other hand, the wretched patients suffered from constant suffocation as well as from a steady aggravation of the symptoms of the disease. This suffocation, the mere result of the want of pure air, was called "dyspnœa"—which is merely Greek for "difficult breathing"—and was treated by opiates and sedatives instead of by opening the windows.

All my consumptive patients, whatever the stage of the disease, live night and day in a pure atmosphere, obtained by allowing a current of air to pass constantly through the room, either by a more or less open window and open fireplace, or by a door opening on a well-ventilated staircase, if the weather does not admit of the window being even slightly open. Rational, reasonable ventilation is not encompassed, however, without trouble and discrimination, for human beings any more than for plants, although it is to be accomplished. Since I have been an invalid I have devoted much time and study to horti-

culture, and have had former convictions as to the
necessity of efficient ventilation thereby confirmed.
Plants under glass, too crowded and not well ventilated, soon sicken, wither, and die. To well ventilate
them, enough and not too much, requires constant
trouble, attention, and good sense on the part of the
gardener; in a word, the exercise of both judgment
and discrimination. Plants that do not get enough
air in glass-houses soon wither and die, and the gardener is discharged as incompetent. Human plants
under the same circumstances wither and die, but
the cause is not recognised, for often the doctor and
his patients are equally in the dark.

Consumptive patients bear ventilating perfectly
well, as well as healthy people, day and night. They
neither get pleurisy nor pneumonia, nor are their
coughs aggravated by breathing pure cool atmospheric air night and day. Whereas all these evils
pursue those who are shut up, as are the numerous
continental patients whom I see in consultation with
their own doctors every winter at Mentone. Moreover, suffocation, medically called dyspnœa, is all but
unknown, even in the latter stages of the disease, to
those who are allowed plenty of fresh cool air. It
soon comes on, however, if the window is shut and
the room becomes close. Persons accustomed to free
ventilation *will* have more air. Indeed, I often stand
aghast at the amount of ventilation such patients,
previously freed by me from groundless fears, insist
on having. All who have damaged lung-tissue, unless

accustomed by long habit to a close atmosphere, feel more or less oppression in a confined atmosphere, owing to the diminished field of their respiration. A concert-room, a theatre, a close chamber at night, bring on dyspnœa all but immediately. This I have learnt from personal experience, and thus most fully can I sympathize with my patients. I fully admit, however, that free ventilation without dangerous draughts is difficult to attain, and that it is much more easily and safely accomplished in a southern than in a northern climate.

Very few persons seem to know, and most of those who do know seem to entirely forget, what respiration really is and means. We inspire and expire, say on an average twenty times in a minute, which is 1200 inspirations in an hour, and 28,000 times in the twenty-four hours. Thus two persons sleeping for ten hours in a room with the doors and windows closed, make between them more than 20,000 inspirations and expirations. At each average inspiration and expiration some twenty-five cubic inches of atmospheric air are taken into and emitted from the lungs. At each inspiration we extract oxygen from the atmosphere; at each expiration we throw out carbonic acid and other offensive products of decay. According to Lavoisier and Sir H. Davy the quantity of carbonic acid gas breathed by a healthy man is 13,453 inches, or about 636 grains of carbon per hour, or eight ounces in the twenty-four hours! Think of the condition of a moderate-sized room

inhabited during the night by two persons, the doors
and windows shut, as is usually the case, after these
20,000 inspirations have been taken and nearly 13,450
cubic inches of poisonous gas have been thrown into
its atmosphere!

Is it surprising that a person coming suddenly
from the pure outside air into such a room should
find it close—that is, offensive from the emanations
of those who have slept in it ? Is it surprising that
the inhabitants of such a room should awake languid
and with a headache, when they have been poisoning
themselves all night with their own foul emanations,
breathed over and over again ? Half a dozen men
sleeping in such a room, or in a close tent, constantly
breed a pestilence, as is seen in lodging-houses and
in camps. Is there not, also, something absolutely
revolting to reflect that in a crowded room, a *close*
room, as it is called, every mouthful of air you
breathe has been many times already down into the
innermost recesses of the bodies of a score or more
people suffering possibly from all kinds of diseases ?

In a really valuable contribution to medical literature by Dr. MacCormack, of Belfast, entitled " Consumption from pre-breathed Air," there is a quaint
conceit which well deserves repeating. " If," says
Dr. MacCormack, " in expiring we threw into the
atmosphere carbon, or visible smoke, instead of invisible carbonic acid, we should really appreciate the
noxious character of pre-breathed air—that is, air
that has been already breathed by other human beings

or animals, and which besides being minus the due
proportion of oxygen, the pabulum of life, is loaded
with the products of animal decay." Dr. Mac-
Cormack's work is the most eloquent pleading in
favour of pure air and physiological respiration that
I have ever read, and the physiological and pathological facts that he brings to the support of his
views are unanswerable. His work deserves attentive
perusal, and would have had more weight with the
profession as to the causation of Pulmonary Phthisis
had he not been too one-sided when he attributes
the origin of this disease all but solely to defective
respiration, to the breathing of pre-breathed air in
close unventilated rooms. I have no doubt that he
is right in very many instances ; indeed, I believe
with him, that it is by far the most frequent tangible
cause. But when he all but discards hereditary
debility, lowered vitality, from whatever origin, as
causes of pulmonary consumption, and when he
thinks free ventilation all but the sole remedy, then
I am obliged to differ.

The carbonic acid which animals evolve during
respiration is heavier than atmospheric air—so much
so that it can be poured over and over again from
one tumbler to another, like water. Thus it falls to
the ground, through specific gravity, and, in a large
assembly concert or ball room, theatre, or church, an
accumulation gradually takes place on the floor and in
the lower atmospheric strata. Under its poisonous
influences nearly all feel oppressed, and the weakest,

delicate children and young women, occasionally become faint and lose consciousness, "from closeness and heat," as the phrase is. They are taken out, and soon recover in the "cool air." The fact is, that they drop down unconscious owing to their being positively poisoned by a foul atmosphere loaded with carbonic acid, and recover on breathing pure air. No degree of mere heat would produce the same result, not ninety or a hundred degrees Fahr.

Those who thus faint in crowded assemblies may be compared to the dogs exhibited near Naples at the well-known Grotto del Cane. This grotto constantly evolves carbonic acid, and dogs are kept to exhibit its influence. A dog is taken by a guide and held over the mouth of the grotto for a minute or two; when withdrawn it appears dead, but after lying inanimate a minute or two it breathes visibly and gradually resuscitates, just like the young lady carried out of church. I, myself, on one occasion, became oppressed and faint when all but alone in the immense but unventilated cathedral of Milan, no doubt from the stratum of carbonic acid lying on its floor.

I remember reading a few years ago an account of a colliery accident, in which the shaft fell in, and 120 men and boys, who were in the mine, died from the foul air before they could be dug out. Some of the miners were educated men, and kept a diary for several days, until they died. This diary was found on them, and from it was learnt that the boys died

first, then the more delicate or sickly men, whilst
the strong men lasted the longest; that is, resisted
the longest the air loaded with the coal gases. Such
is our church and concert-room experience. The
weakest succumb, are poisoned the first, but all must
and do feel its influence.

Nothing surprises me more every winter at Mentone than the perfectly insane manner in which the
Germans and the natives of Eastern Europe, patients
and physicians, with all their profound knowledge,
treat the question of ventilation. It would seem as
if everything they learnt in their scientific books and
in their Universities about the physiology of respiration was forgotten and ignored as soon as they entered
into public life. I see a good deal of German practice, for since my work on "The Winter Climate of
the Mediterranean" was translated into German,
Germans have come in shoals in winter to Mentone.
In the year 1869–70 we had above four hundred,
with several able German physicians.

The first thing a German or East of Europe invalid
does is to order a stove, formerly unknown in the
place. The pipe is carried into the chimney through
the fire-place, and the latter is hermetically closed
up, plastered around the pipe. The latter has a
key, which is turned as soon as the wood burnt is
reduced to embers, so as to confine " the heat" or to
throw it down into the room. Thus is closed the
only communication with the exterior. Still further
to shut himself out from his real friend but supposed

enemy, the poor German or Russian invalid has canvas piping put round the windows and doors, which are constantly shut. To crown the whole, a thermometer is placed on the wall, and the temperature kept day and night at 65° or 70°. This is constantly done with the approbation and connivance of the really learned German physicians who attend him. So he stews and perspires in the day, and has cold sweats at night, suffers from dyspnœa, which means in reality that he is suffocated, and cannot breathe (no wonder, poor fellow !) To remedy the dyspnœa (suffocation) he is drugged with morphia and codeine, until his pulmonary nerves are so paralysed that they do not rebel against foul air and half-oxygenated blood. Then, if a cold wind comes, as it does occasionally come everywhere in winter in temperate climes, as he cannot always stop in this heated poisonous box, he goes out, and often gets pleurisy or pneumonia, laryngitis, pharyngitis, bronchitis, *et hoc genus omne.* Whilst my patients, equally ill. but living in pure air, with their windows or doors open often day and night, very rarely indeed suffer from such complaints.

It is impossible to travel in Germany and in Eastern Europe generally without everywhere observing this ignorance, this utter contempt of the physiology of respiration. Everywhere in the large hotels we find double windows, stoves fed from the outside passages, and every possible contrivance resorted to in order actually to prevent the renewal of the air in inhabited rooms. The influence of such oblivion of physiology on the part of the medical profession must be very fatal, and must lead, as Dr. MacCormack so eloquently argues, to much preventible disease. I cannot but make the medical profession in these countries responsible for the state of things, for they could alter it if they would. But they live and die like their parents, victims to their neglect of physiological laws. I have seen several estimable and learned consumptive German physicians die under my eyes at Mentone in the poisonous atmosphere above described.

The principal cause of this neglect of ventilation in health and disease, is no doubt to be found in the extreme severity of the climate in Eastern Europe. In a region where rivers are frozen several feet deep and snow lies on the ground for months, where, also, fuel is scarce and dear, it is quite comprehensible that the cold atmospheric air in winter should have been considered an enemy, and should have been carefully excluded, as long as its chemical composition and as the nature of the respiratory processes were not understood. Although such ignorance no longer exists, the prejudices and habits of centuries of ignorance still hold their usual sway.

It must also be acknowledged that in a very cold climate it is extremely difficult to renew the air and yet to keep it at a safe temperature for habitual respiration, say at or about 50° Fahr. My complaint, however, is that the necessity is not recognised, and that no attempt is usually made to meet the difficulty.

I would draw earnest attention to the physiological fact that the lungs are made to breathe cold as well as warm air—indeed, air of any temperature from zero to 100° Fahr.—just as the face is made to bear exposure to the external atmosphere. How could the lungs be protected, if they required protection? which they do not. Domestic animals that live out in the open air winter and summer are freer from colds than those that are coddled in warm stables. Men who are much exposed, and constantly breathe air at a low temperature, are less liable to colds and influenzas than those who live constantly in warm rooms. All who have horses are aware that to keep a stable warm is the surest way for the inmates to suffer from constant colds.

The efficient ventilation of a room is all but impossible unless either the door or the window be left open, or partially open. All the contrivances hitherto invented to ventilate a closed room are altogether inefficient. They do not supply and renew the air consumed even by one or two individuals, day or night, in an inhabited room, even a large one. In the summer a window may be kept more or less open; in cold foggy weather a door, the staircase and lobbies being well ventilated. In such a room there will be draughts. It is impossible otherwise to thoroughly ventilate without them, but the inhabitants can avoid them—that is, by getting out of their way.

I may mention two facts that aptly illustrate the evils of defective ventilation. Some years ago I was riding in the Highlands of Scotland with a local proprietor, when we came upon a village of well-built stone houses with slate roofs, which strongly contrasted with the miserable shanties or hovels generally met with. On my complimenting him on his rebuilt village, he told me that he had acted for the best in erecting these good weather-proof houses for his tenants, but that, singular to relate, they had proved more unhealthy than the miserable dwellings which their occupants previously inhabited. Fever and other diseases were rife among them. On close examination I found that there were double doors, and that the windows were fastened, and never opened. I have no doubt that their comparative unhealthiness was in reality owing to their being quite weather-tight, and consequently unventilated. In the miserable hovels they previously inhabited, if the rain of heaven came in, so did pure air.

The other fact is narrated by Professor Hind in a recent interesting work on Labrador. Consumption appears to be all but unknown to the natives living wild in the fastnesses of this desolate region, in tents made of spruce branches imperfectly lined with skins, and more or less open on all sides to the external air, although they are exposed to famine and every species of hardship. But when these same natives come down to the St. Lawrence to take a part in the fisheries, occupy well-built houses, and, being well paid, live

E

in comparative luxury, many of them, in the course of a year or two, become consumptive and die miserably. I am fully impressed with the idea that the development of the disease under these circumstances is principally the result of their living in closed houses in a vitiated atmosphere, as is no doubt the case in our own towns.

In all the statistics of the mortality from Phthisis that have been published there is a steady and invariable relation between the number of deaths from Phthisis and sedentary or out-door life. This is observed in all climates, in all latitudes, at all elevations. Town life, manufacturing, mining, sedentary in-door pursuits of any description, moisture and dampness of climate, which drive people indoors, increase the mortality. Out-door pursuits, agricultural life, decrease it. There is no better established fact in pathology.

In the south of Europe there is an all but general belief in the contagion or communicability of Phthisis, and this belief is, no doubt, founded on the misinterpretation of facts actually observed. They shut up their consumptives, breathe the pestilential air, and often become consumptive themselves. I have myself, in several instances, known the relatives of consumptives, who have lived with them and nursed them, become attacked with Phthisis. I have, however, always been able to trace the origin of the disease to co-habitation in an imperfectly ventilated room, which alone will produce Phthisis in healthy individuals, apart from contact with phthisical persons, and without appealing to contagion or communicability.

The following case admits of the former interpretation :—A strong, healthy, well-made husband, aged twenty-seven, with no hereditary or constitutional taint or weakness, came over from Australia—a four months' journey—in the same cabin as his wife, who was in the last stage of suppurative Phthisis. She died soon after her arrival in England, and he came to Mentone that winter a confirmed consumptive, dying himself subsequently. He was perfectly well when he stepped on board the vessel in Australia ; but in a small confined cabin breathed for months an atmosphere loaded with pus particles thrown out of the suppurating cavities of his wife's lungs, possibly to his own destruction. The disease may, however, have been owing merely to bad air and anxiety.

Such cases are probably not unfrequent in countries where consumptives are still closed up, and live with their near relatives in all but hermetically sealed rooms, and they explain the southern belief in the contagion of Phthisis. It is this belief that leads them to destroy bedding, clothing, furniture, after death from Phthisis. The recent researches and experiments as to inoculability of the secretions from tubercular and caseous lung-deposits give, at first sight, some plausibility to these views.

On injecting into any part of the body tubercular or caseous matter from the lungs, or the purulent contents of phthisical lung-cavities, it has been superabundantly proved that deposits of a similar character—tubercular, or caseous, or purulent—are formed in the lungs and elsewhere. Such a result, however, it must be remembered, often follows the injection of purulent matter of any kind taken from any part of the body. This fact once accepted, I do not see how we can reasonably repudiate the possible inoculation by the lungs of these same morbid products, inspired as a kind of purulent spray in a badly ventilated room for weeks and months, day and night.

Some interesting experiments made on dogs by Dr. P. Buhl at Munich, and quoted by Dr. Thaon of Nice (*Nice Médicale*, January 1, 1878), have a direct bearing on this subject. Three healthy, vigorous dogs were selected, and confined in a kennel closed on all sides, except the door, which was fenestrated with wire. Through this wire opening was introduced daily for six weeks, during an hour and a half, with a spray-producer, air loaded with the expectoration of a consumptive presenting a large cavity. Two other dogs had two table-spoonsful of phthisical sputa mixed with their soup daily. At the end of six weeks they were killed, and the lungs were found to be the seat of general miliary tuberculisation, although the dogs had not presented any symptoms or constitutional disturbance. Animals, such as rabbits, dogs, cows, horses, are very susceptible to purulent inoculation in any shape.

There is no fact, however, better established than

that Phthisis is not contagious, like scarlet fever or
smallpox, or infectious like typhoid fever. Were it
otherwise, we should have daily evidence of the fact.
But in the face of the results that these researches
have brought to light, it seems to me impossible to
deny that it *may* be communicated to the healthy by
breathing constantly air saturated with the purulent
secretions of advanced Phthisis. This is an all-
powerful argument for the free ventilation of rooms
occupied by the consumptive, for the sake of those
who attend them and live with them, as well as for
their own. In a confined atmosphere they probably
poison themselves by their own fœtid breath, and
extend disease to the healthy regions of the lung.

Having given this solemn warning, I must add,
most emphatically, that I have never once seen
a case that could lead to the suspicion of communication when the rooms inhabited by the invalid have
been well and freely and constantly ventilated.
Under such conditions, I think no one need fear to
live with a consumptive relative, friend, or patient.

Attention to the functions of the skin may be considered next in importance to attention to food and
air—that is, to digestion and to respiratory nutrition.
The skin has very important eliminatory functions to
perform. It is by excretion through its pores that
the economy partly throws off the effete or used-up
carbonaceous and nitrogenous elements of the system.
This is illustrated by the strong odour of the cutaneous secretion when not washed off. Moreover, the

skin and the lungs seem to partly replace each other in this work of excretory purification. In warm summer weather the skin and liver act freely, and the lungs and kidneys are comparatively at rest. In the cold damp weather of winter the pores of the skin are closed, and it rests partially, as does the liver, the lungs and the kidneys taking up the excretory processes. Thence, probably, the feverish colds of cold damp weather. The blood is poisoned with the elements that the closed pores of the skin should have eliminated, which occasions the fever; whilst the lungs often succumb to the increased duties they have to perform, inflammatory affections supervening. Whatever the explanation, the fact is certain, and it is now well established, that one of the best modes of preserving the respiratory organs from winter cold is to keep the pores of the skin open by the use of cold or tepid water, combined with friction ; or, in other words, to keep the cutaneous excretions up to their normal standard.

Acting in accordance with this view, I make all my consumptive patients, whatever their condition, if they have the strength, use a sponge-bath at a temperature of from 65° to 70° daily, and with the greatest possible benefit. I neither have to contend with hæmorrhage nor chills, nor with aggravation of the cough, but quite the contrary. The cold sponge-bath produces, in nearly every instance, a feeling of indescribable comfort, and lowers the pulse. The contact of the cold water may, just at first, accelerate the expectoration of the muco-pus collected during the night in the bronchial tubes, but that never alarms when it is explained that such a result is naturally to be expected. I myself derived the greatest possible comfort and benefit from cold sponging in summer, in the open air, on the banks of Loch Awe in Scotland, the waters of which were at 62°, and that when I was very ill, pulse above 100, skin hot and feverish. I committed this, as I thought, rash act in a fit of desperation, but to my delight and surprise I felt immediately better, the pulse falling and the feeling of burning heat diminishing. This gave me a confidence I have never lost, and of which I have never had reason to repent.

I have followed the practice above detailed for so many years with such a number of consumptive patients—for it is the rule with me—that I can speak authoritatively not only as to the all but constant advantage to be derived from it, but also as to its freedom from risk or danger. A case that occurred, however, soon after I commenced cold sponging in advanced Phthisis shows how the judgment may be warped by unfortunate coincidences in all such questions.

A young gentlemen of twenty was placed under my care at Mentone. He had phthisical deposits in the upper portion of both lungs, and a small cavity 'at the apex of one, and was otherwise very ill. Under the influence, however, of climate and treatment he made a rapid rally, and in the course of the first three months gained ten or twelve pounds in weight, so that we were all rejoicing and hopeful. He used the cold sponge bath daily on rising with apparent benefit and pleasure. One morning he got up as usual, prepared for his bath, and had raised one foot from the ground, when he coughed and brought up some blood. He immediately returned to bed and sent for me; I was living in the same hotel, and was with him in a few minutes. I did all I could, but the hæmorrhage was irrestrainable; if checked for a few hours it returned again, and he died on the third day. Evidently a large vessel had given way in the walls of the cavity or near it. Now, had the hæmorrhage commenced half a minute later, whilst in the bath, he would have thought that it was the cold water that had occasioned it, and so probably should I. *Post hoc propter hoc.* Thus from a pure coincidence I might have been led to think cold ablutions precarious, nay dangerous, and might have given them up, or resorted to them in fear and trembling.

My clinical rule is to begin with water about 70° Fahr., descending or ascending until I reach the temperature suited to the individual case as proved by healthy reaction. When the patient is too weak for the bath, the body should be daily sponged, in part or entirely, whilst in bed.

The question of exercise is an important one, and one that requires discussion and elucidation. From personal experience and observation, I believe it is a great mistake for consumptive patients, debilitated by disease, weak, below par, to take much active exercise. Every winter I see some such patients walk themselves to death. They have been told by their medical attendants at home to take exercise, and they do so, thinking that what gave them an appetite and did them good when well, will do so now they are ill; but they merely walk themselves into their graves. The disease from which they are suffering is one of debility. The strength of former days has gone out of the youth and out of the man, although perhaps he knows it not. Or the strength he has is fictitious, unreal strength, the result of a febrile condition, of a state of morbid nervous excitement. So, having nothing else to do, he walks from breakfast to dinner, up hill and down dale, loses his appetite, cannot eat, becomes "bilious," is dosed for liver, and the disease progresses rapidly. Every winter, towards January or February, some invalids consult me who have up to that time taken their case in their own hands, and have thus walked with the healthy, in order to gain strength. But they have lost it instead—have become paler and thinner ; and when I see them, I find that they have lost ground, that the disease has gained upon them since they arrived in the autumn, and that they are decidedly worse—all from over-exercise.

All human effort or work, muscular, nervous, or organic, is attended with the expenditure of force.

To be spent it must be there, just as for a man to spend money he must have it. To demand of a weak debilitated invalid to spend in muscular exercise, strength, power, that he has not got, that his imperfectly digested food does not supply him with, is unphysiological on the one hand, thoughtless and cruel on the other.

The sound rule for a consumptive patient is to be as much in the open air as possible, but to take passive exercise, not active; to ride in an open carriage, to be rowed in a boat, to sit and lie hours in the air, to live with windows open, but not to incur great muscular exertion. The amount of vital power in such cases is small. If it is too freely expended in exercise, there is not enough left for normal digestion; food is imperfectly assimilated, nutrition is defective, and the disease progresses.

A singular but explicable fact is, that during the existence of active disease, when morbid deposits are forming and softening, very often no lassitude is felt on exertion. But when the disease is arrested, and a curative process has been set up, extreme debility and lassitude may be experienced and complained of, lasting for months or even years. I felt this lassitude for many years to a painful, extreme degree, and could scarcely persuade myself that it was not really a bad symptom. The explanation is, however, very clear and simple. In chronic as in active disease there is often a false, feverish strength, like that of the delirious patient whom it takes half a dozen men to hold. In the curative convalescent stage in either case, the false strength is gone ; the real condition of the sufferer comes to light, as it does with the delirious patient when the delirium is gone, and he can scarcely lift his hand from the bed. So the convalescent phthisical sufferer may feel much weaker than when he was really ill, and scarcely believes the flattering statement of his physician or the testimony of his own senses.

The social and mental hygienic conditions favourable to the treatment of Consumption may be summed up in a few words—rest, repose, the absence of the ordinary duties, cares, harass, and worries of life, the control of the passions. To obtain these is difficult in the social medium in which the disease has appeared. Therefore, the duties and obligations of life should be surrendered for a time, if possible, modified, diminished, if not. Those, however, have the best chance of arresting the progress of disease who can escape from the social medium in which it appeared, in which there are often causes of distress, of mental agony, which are not revealed to the physician, or perhaps even to relations or friends. To do this it is always necessary to make great sacrifices—sacrifices which many cannot make. Those who can should remember that the struggle is one, not merely for a higher or lower stage of health, but for life itself. Would that they could see it in this light ! but few will consent to more than a very temporary or partial surrender of position, income, duties, ties, fame. They shut their eyes, and often die when they might have been saved.

Every year I see illustrations of this deplorable want of courage to face the impending danger, to struggle manfully with death. I must add that the mistaken kindness of physicians who will not tell their patients the truth for fear of distressing them, and the new phraseology of disease introduced and adopted by the disciples of Virchow, contribute to diminish the chances of recovery from Pulmonary Consumption. How can patients struggle against a dangerous disease, fight the battle of life, if they are not aware of its existence, if they think they have merely limited pneumonia, a patch of caseous pneumonia, a catarrhal or scrofulous deposit, terms under which this fell malady, which unchecked implies death, is now so often concealed ?

I constantly see cases in which persons suffering from Phthisis could make the necessary sacrifices for treatment if they would, but they cannot bring themselves to do it. They have ties, duties that must be attended to ; they want six months—a year or two, and then they will do all they are asked. Such instances crowd on my memory, but I will only narrate the two following.

Some years ago an eminent railway contractor called on me in London in July for my opinion on his case. He was a strong, well-built man of fifty, and had had a cough and other thoracic symptoms for a year. On examining the chest I found a well-marked phthisical deposit in the upper part of the left lung. There was dulness for two inches below the clavicle, and also posteriorly in the supra-spinata fossa, bronchial respiration, mucous rattles, no cavity. The other lung was apparently sound, the constitutional symptoms not severe. He had never given up work, was not aware of the serious character of his disease, and there was no known hereditary tendency to Phthisis.

The case was clearly a good one for treatment; and the disease was accidental, limited, and the constitutional powers were, so far, only damaged—not wrecked. I advised him to give up his professional duties, which he told me were very heavy, and to migrate to the south of Europe for the winter months. This he said he could not do that winter, but would do the next; it would necessitate his giving up all his contracts, and if he did so he should lose a hundred thousand pounds. My reply was, " If you can afford to lose it, if losing it still leaves you sufficient means, make the sacrifice, and fight the battle for life. It is not absolutely necessary to go south—you could go to Torquay, Ventnor, or Bournemouth, but if you do you will no doubt keep your contracts, your anxieties, the mental labours and trials which have brought you to your present state, and you will probably get worse instead of better." He left me undecided, kept his contracts, died the following summer, and left three hundred thousand pounds to two daughters, His clinging to the one hundred thousand depending on the contracts probably cost him his life.

Again, a few years ago a City merchant called on me with his medical attendant, nearly in the same condition. He was fifty-two, childless and prosperous, as he told me himself. I gave him the same advice, but saw and heard no more of him until two years after the London consultation. On coming into my study at Mentone he reminded me of his London visit, adding, " My call upon you that day has cost me more than ten thousand pounds. I have done as you advised me. I have entirely got rid of my business, losing that amount of money in the process, and now here I am." But, in the meanwhile, during these two years, the disease had advanced, and the chances of recovery were gone, very different to what they were when I first saw him. He died within the year.

One more lesson from my note-book. Soon after the Suez Canal had been opened a large English steamer, built principally for commercial purposes, passed through it, and steamed successfully to Ceylon in fine weather. There she took in cargo and two hundred passengers for the return voyage. But the monsoon had commenced, the sea was rough, winds adverse, and the engines proved too weak and in-efficient. Instead of steaming ten knots an hour she scarcely made four. The coal was at last exhausted, and the vessel would have lain at the mercy of the waves had not the cargo been of a consumable nature —viz., coffee and cinnamon. One thousand pounds worth of these precious commodities were consumed daily for twenty days, but on the twentieth day the steamer reached Aden, at a cost of twenty thousand pounds in fuel. But the passengers and crew were saved. Sometimes we are called upon to imitate the Ceylon steamer, to burn coffee and cinnamon to save the human ship from destruction. This tale was told me by one of the passengers.

Even the healthy must control their passions if they wish to retain health, but it is still more necessary for those who have lost it to exercise this control. Often those who appear calm and self-possessed are, in reality, a prey to conflicting emotions and passions. Such a mental condition is a positive impediment to the recovery of health in chronic disease. This I have frequently observed often in cases too dramatic to be here narrated.

CHAPTER III.

IF, as I have assumed, the deposits in the lungs which
constitute Pulmonary Consumption are a disease of
defective nutrition, itself the result of exhausted or
lowered vitality, the debated question as to what
climate is the most calculated to arrest and cure the
disease, is easily answered.

Theoretically, or rather physiologically, a cool, dry,
sunny, stimulating climate is the one most likely to
rouse depressed vitality and health ; not a warm,
moist one, or an exceptionally cold one. Practically,
my own experience and that of many other observers
show that such is the case, that consumptive patients
do best in a dry, cool, sunny region, that they are
rather damaged than improved by a warm, moist climate, and that they are exposed to great risks in
very cold ones.

This, the modern view, is certainly one of the most
valuable contributions that modern science has made
to the treatment of Phthisis. And yet, although
most true, I cannot say that it is a principle of treat-

ment generally understood or practically carried out
even in our own country, which is, I consider, much
in advance of the Continent in the rational treatment
of Consumption. Indeed, abroad the hygienic and
climatic treatment of Phthisis is still in its infancy,
and nearly all the old errors are in full operation.
That such is the case is clinically demonstrated
by the numerous cases of Phthisis that 1 see
every winter at Mentone from central and eastern
Europe.

Forty years ago, when Pulmonary Consumption was
generally considered to be entirely an inflammatory
disease, when undue importance was attached to the
inflammatory conditions—bronchitis, local pleurisy,
local pneumonia—which characterise its various
stages, it was quite natural that warm weather and
warm climates should be considered desirable. In
warm weather, as we have seen, when the skin and
liver are acting vigorously in the work of blood purification, inflammatory affections of the lungs are
neither common nor severe, and are easily subdued.
Was it extraordinary that a warm climate should be
thought the one thing desirable in the treatment of a
disease the prominent outward symptoms and features
of which are these very inflammatory affections?
And yet the results obtained by the antiphlogistic
or lowering mode of treatment then adopted for all
inflammatory diseases, acute or chronic, combined
with warm air, were so little satisfactory that most,
nearly all, phthisical patients died, and the disease

F

itself obtained the reputation of being all but incurable.

As I have shown in the first chapter, when discussing the nature and causes of Pulmonary Consumption, modern histological research has led many pathologists back to the Broussaisian doctrine of inflammation, as the *fons et origo mali*. According to these doctrines catarrh and pneumonia mostly initiate the lung deposits which, clinically, constitute Phthisis. Were we still under the influence of Broussaisian therapeutical doctrines, all who adopt these views would necessarily revert to bleeding, leeching, and warm climates, as the remedies indicated. But the therapeutical ideas connected with Broussaism are no longer the order of the day. We have gone back to vitalism, which is a real progress, provided the vitalism adopted be rational, not carried too far, or erroneously interpreted to mean mere alcoholism.

Although professing the inflammatory origin of the lung deposits in Phthisis, Virchow, I believe, is ready to admit that a depressed or lowered state of vitality is at the root of these low forms of inflammation. This fact his followers should always remember, although they often do not remember it. Professor Bennett, of Edinburgh, whose authority as a histologist is equal to that of Virchow, considering tubercle to be, undeniably, a disease of defective nutrition and debility, and looking upon the lung deposits as generally tubercular, is most urgent in recommending sthenic measures of treatment. Thus, whichever of these great authorities we follow in our doctrinal views of the origin and nature of the disease, clinically, therapeutically, the indication is the same—to strengthen, to invigorate by every means in our power.

If, therefore, the inflammatory conditions of the lungs, in Phthisis, whether primary or secondary, are merely epiphenomena, symptoms of organic debility, of exhausted vitality, it is clear that warm weather cannot cure them. Indeed it is much more likely to aggravate them, by increasing the organic debility which is at the root of the evil, and by inter-. fering with the restoration of healthy nutrition which alone can arrest their progress.

WARM CLIMATES.

Practical experience proves that such is the case, that warm climates accelerate the progress of Phthisis. Such is the experience of the medical profession in many different climes and regions. Perhaps the most valuable and conclusive evidence on this subject is that furnished by the English and French Army Reports during the last forty years. From them it has been established that soldiers suffering under Pulmonary Consumption get worse in all warm climates, especially during the summer—in the East and West Indies, at Malta, in Algeria. Consumptive soldiers are now sent home to a temperate region from all such climates, as the best course that can be followed for their welfare.

I may here mention a valuable illustration of this fact, drawn from private experience. My late valued friend, Dr. Dundas, practised for twenty-three years in Bahia in the Brazils, a tropical climate, leaving in 1843. Many years ago he told me that during his residence there he was constantly receiving patients from Europe affected with Phthisis, sent by the profession who then relied on the influence of a warm climate, as to one of the best regions that could be found for their disease. These patients invariably got worse, and died much more rapidly than if they had remained at home. So fully did he become impressed with the conviction that the climate—a very healthy one in all respects apart from its tropical character—was deadly to the consumptive, that if any of his own patients among the European population were thus attacked, he instantly sent them back to Europe.

These facts are at once explained when we recognise the principle that at the root of Pulmonary Consumption there is debility, exhausted vitality, and not merely inflammatory disease of the respiratory organs; that it is an affection in which the indication for treatment is to strengthen, to invigorate, not to soothe and calm " symptoms." Warm weather produces languor, a disinclination to take exercise and to eat, often a positive disgust for meat and for fatty substances, and interferes with sound sleep. In warm weather the natural desire is to remain recumbent, idle, and half-dressed, to drink lemonade, and to eat ices. The attempt to take a fair amount of nitrogenous and carbonaceous food, as a duty, is often followed by disturbed conditions of the liver and of the digestive and intestinal organs generally, to which there is so great a tendency at all seasons in the consumptive. I only ask, in common sense, is such a state of things, is a temperature or a climate that produces such results, to be relied on in the treatment of a disease of debility ?

TEMPERATE CLIMATES.

A temperate, cool climate, on the contrary, with the thermometer varying between 55° and 65° Fahr. in the day, and between 45° and 55° in the night, has a very different physiological effect on the constitution. It braces and invigorates the system, it stimulates to exercise, improves the appetite, and admits of the ingestion and digestion of both meat and fats, so necessary for perfect nutrition. Thus it favours our attempts to rouse vitality by improving the nutritive functions.

The temperature which I describe, one ranging from 45° Fahr. at night to 65° in the daytime, is physiologically the most conducive to the well-being and longevity of the human race. The extremes of cold and the extremes of heat, on the contrary, are not conducive to longevity. In this, as in everything else, we may say with the Latin poet, "Medio tutissimus ibis." In warm climates generations succeed each other more rapidly than in temperate ones.

The inhabitants marry early, reproduce their race early, inherit property and arrive at positions of trust early, and die early, to make room for the next generation. This is the case in India, and in tropical countries in general. In temperate regions the span of life is longer, the succession of its phases is less rapid, less feverish. Thus Scotland, essentially a cool, temperate climate, is also one of the healthiest countries in the world ; the average duration of life being, I believe, above that of any country in Europe. It is certainly above that of England, where it is above the average of continental Europe.

That a medium temperature, one in which the thermometer does not descend below 45° **Fahr.** at night, or ascend above 65° in the day, is the most conducive to human health and to longevity is proved by the Registrar-General's reports in the British Isles, and by the mortality register all over the world. In the British Islands the healthiest summers are those that are lowest in temperature ; the healthiest winters those that are highest. The seasons of least mortality in the year are those in which the temperature is neither extreme in one sense nor in the other. Conversely, the years and seasons of greatest mortality are those in which extremes of cold in the winter, and of heat in the summer, are reached. In warm climates, the deadliest seasons are those of greatest heat ; in temperate climates, the deadliest seasons are those of greatest cold.

COLD CLIMATES.

The above generally recognised facts are consistent with and explained by physiology. Extreme
heat and extreme cold not only interfere with the
equilibrium of functional activity, throwing a strain
on some of the functions of animal life, to their
serious risk and danger, but necessitate modes of
existence detrimental to the healthy performance of
these functions. Thus, in very cold climates, such as
St. Moritz in the Engadine, and St. Paul in Minnesota, United States, which have been recommended
of late for Phthisis in winter as well as summer,
invalids have to live in winter for by far the greater
part of the twenty-four hours in badly-ventilated
rooms. When they go out they have to undergo
the transition to a temperature thirty or forty, or
even more, degrees less than that in which they live
during the greater part of the twenty-four hours.
Such confinement, such transitions, even much less
marked ones, constantly give rise, in all northern
countries, in winter, to inflammatory affections of the
mucous membranes of the aerial passages, to pneumonia, and to pleurisy, and that in the healthiest
members of the community. If it is so with the
healthy, how can we expect those to resist such
influences who are already diseased, who have morbid
deposits, inflammatory, catarrhal, scrofulous, tubercular, in their lungs, softened or not? How can

those who have already local pneumonias, local
pleurisies, expect to withstand their pernicious
influence ?

Moreover, they do not even live in the pure
mountain air which they have ascended five or six
thousand feet to reach, but in an atmosphere
vitiated by stove-heating, by their own respiration
and by that of their companions. Even at Clarens
and Montreux, on the lake of Geneva, I found the
houses and passages heated in winter by large stoves,
and the landlords and their tenants had strict
injunctions to keep the temperature night and day
up to a given standard.

Nor do either the sick or the well escape the
ordinary influences of cold and damp, only it is difficult to get at the truth in such questions. The fact
of a few chest invalids passing scathless through a
winter in such extremely cold, or cold and elevated
regions, or even of their benefiting in health thereby,
does not prove that they have not run great dangers,
or that their experience is that of their companions.
A person may live many years in good health at
Sierra Leone, but his escaping illness and death does
not prove that it is a healthy locality. Nor does a
man escaping from a bloody battle prove that he ran
no danger in the action. We must remember that
the grave is very silent. Those who succumb to
pernicious influences or to danger, once buried, do
not return to say what has occurred. Again, statis-

tics are very delusive in such matters. There are waves, tides, in medical matters, in chronic diseases, as in other sublunary events. I see this at Mentone, as it is seen elsewhere. Some winters many bad cases come to me—cases in which the lung is like an empty decanter, reduced to its walls—cases with incurable liver or kidney complications—and I lose many patients. Perhaps the next winter most of the cases that present themselves are favourable ones, in the incipient stage of disease, and free from unfavourable complications, and consequently but few die. How is it possible, in reason, to compare these winters, to draw averages or statistical data founded on them?

In my opinion it is safer to be guided by our laboriously-acquired knowledge of physiology and pathology; to believe that their laws are not suspended, either in winter or in summer, at the summit of the Engadine mountains, at St. Moritz 6100 feet above the sea level, or on the plains of Minnesota, where the lakes and rivers freeze many feet deep. The probability, nay the certainty, is, that were the entire truth known it would be found that the influences produced by extreme cold in London, Paris, St. Petersburg, New York, are also produced in these regions, and that to consumptive invalids they are dangerous, to say the least. A case which I will briefly narrate illustrates the risk incurred by despising, setting at nought, the rules that guide the profession in the prevention and treatment of disease.

An American gentleman of fortune, aged forty, with incipient Phthisis, was advised by his physicians at New York to pass the winter of 1867 at Mentone under my care. On arrival in Europe he met with some friends who were going to Pau, and who asked him to accompany them. This he did, but instead of placing himself under one of the well-informed physicians who are there practising, he took his case in his own hands. When he was well hunting had done him good, so he determined to hunt, and went out with the hounds two or three times a week. When spring arrived he felt worse, and came over to me at Mentone. I found a decided morbid deposit at the apex of the right lung, with softening; but the disease was localized, and limited to the upper region of the one lung. The general health had run down during the winter, owing no doubt partly to the violent exercise he had taken and partly to the utter want of proper management and treatment. We agreed that he should pass the summer at Dieppe or at Ostend, and that the following winter should be spent at Mentone. The next October he made his appearance in my consulting room, and told me that he had travelled about and had passed some months at St. Moritz with much benefit to his health. I found that the lung was more compromised than in the previous April, but still considered the case a favourable and hopeful one for treatment. There was no phthisical disease in the family, and his own health had been good until eighteen months or two years before. His father, a healthy old man of eighty, was with him, as also a family of healthy children. But he was of a restless, undecided nature, and did not settle down. He had been told at St. Moritz that it was a mistake to come south, that sunshine and a mild temperature were quite an error, that what he wanted was "bracing-up" with Engadine cold in winter. He became slightly bilious, walked about moodily, looking at the sunshine and fine weather as enemies that would kill him if he remained. So one day he called, not to consult me about his plans, but to take leave, as he was starting the next morning for St. Moritz.

He was as good as his word, and left us the following morning for Geneva, on his way to the Engadine Mountains. He arrived at Geneva in a snowstorm, on the 20th of November, was taken ill next day with acute pneumonia, and died on the fifth day, just one week from the one on which he took leave of me, tolerably well, with merely a limited amount of chronic disease in one lung. I really believe that this man's life might have been saved had he allowed himself to be guided by me, and not fallen into the St. Moritz "winter delusion."

I look upon the recent movement, which tends to send consumptives, and chronic chest disease generally, to very cold and elevated regions for the winter, as a mere reaction from the therapeutical doctrines of our fathers. In the United States, Florida, and Nassau, which are mild and moist in winter, were, until lately, the resort of such cases. That is, whilst the reigning doctrines demanded warmth and moisture for chronic chest disease, our American brethren sent their patients to Florida and Nassau, as we sent ours to Madeira and to the West Indies. Now that medical doctrines have changed, that vitalistic, sthenic views of treatment prevail, and are found to give infinitely more satisfactory results than those that followed antiphlogistic treatment, the medical mind, in America as in Europe, looks about for a colder climate.

As usual the pendulum has a tendency to pass to the other extreme, to go from Madeira, Jamaica, and Barbadoes, from Havannah, Florida, and Nassau, to the ice-covered summits of the Swiss mountains, to the frozen plains of Northern America. Many minds can never "constitutionally" accept and follow the golden adage which I have taken as my motto, " Medio tutissimus ibis." They cannot remain in the middle of the road, they must pass from one extreme to the other.

Having given the reason which should lead us to avoid extreme climates, extremes of heat or cold, either for summer or winter, in the treatment of pulmonary consumption, we have to consider which are the best "attainable" climates where a moderate or medium temperature is to be found in summer or in winter.

TEMPERATE SUMMER CLIMATES.

We may eliminate at once all tropical regions, all climates in which the mean annual temperature is high, above 60°, or where the winter mean is much above 50°. We must remember that a moderate annual mean may result from great heat in summer, as at Corfu, where the annual mean is 65°, with a summer mean of 77°, and a winter mean of 54°. At Mentone also, the annual mean of 60°·80 is drawn from a winter mean of 49°·5, and a summer mean of 73°. (See Table, p. 621, in my work " Winter on the Mediterranean.")

Firstly, as to summer, there is perhaps no better climate in the world for the consumptive than the British Isles. The nights are generally cool, and the days temperate, the thermometer seldom rising above 70° in the shade. Sometimes, however, it does ; we have " dog days" in England, days when the thermometer reaches 86°, or even ascends above. This degree of heat is very trying in England, even when the nights are cool, on account of the generally moist state of the atmosphere. Owing to our insular position, and to the warm water of the Atlantic and of the Gulf Stream impinging on our western shores, the atmosphere is generally loaded with moisture, and the sky partially covered with clouds even in summer. Warm moisture stops insensible perspiration, and is very oppressive.

The moist heat of midsummer, in the southern

and central parts even of England would, sometimes, be unbearable, were it not for the uniform coolness of the nights. However hot the day in Great Britain, the thermometer is seldom above, generally below, 60° at night. Moreover these periods of extreme day heat seldom last more than a few days. Thus, by free ventilation in the evening, and by opening windows at night, a cool night temperature may be nearly always attained. When such is the case the heat of the day has a much less pernicious and trying effect on the human organization. What makes the heat of continental and southern Europe so difficult to bear, and so pernicious to health, is the fact that the nights being nearly as warm as the days, there is no rest for the organization, and that the duration of this trying day and night heat extends over several months.

There are, however, exceptionally warm summers in England proper in which the thermometer may remain for several weeks above 70° in the daytime. This all but tropical weather is detrimental to consumptives, but they can easily escape from it by going north, to our north-eastern counties, to Wales, to the Highlands of Scotland, or to Ireland. The west coast of Scotland is proverbially moist ; but moisture with the thermometer between 55° and 65°, as it generally is in summer in the Western Highlands, does no harm whatever ; neither causing cold nor cough, nor increasing them if they already exist. The Western Highland and Islands are stated to be exceptionally free from Phthisis. A consumptive person, with a bad cough and free expectoration, provided he be warmly clothed and protected from rain, may sit in a boat all day on a Scotch or Irish loch, exposed to showers, in summer, with the thermometer at about 60°, without taking any harm. I have done so myself for weeks and months together, not only with immunity, but with the greatest possible benefit to the general health, and as a result, to bronchial suffering.

Continental Europe is by no means so suitable as a residence for mid-summer as the British Isles, owing to the great heat which nearly everywhere prevails from June to September. The coasts of Normandy and of Northern Brittany, and those of Holland, however, share with the British Isles the milder temperature which winds from the northern seas, and a canopy of vapour and cloud give us, the latter, protecting us from the rays of the sun. In central and southern Europe, and even in Switzerland, the only means of escaping great and pernicious heat is to ascend the mountains several thousand feet. But in the higher regions great care is required, and there is a serious drawback. The nights are often very cold in fine weather through radiation, and in wet weather you may be in the clouds, that is, in cold vapour and mist, day and night, for days or weeks together.

Within the last few years the mountain regions of Switzerland, and especially the Engadine, or valley

of the Inn, in the canton of the Grisons, eastern
Switzerland, have attracted much attention, both as
resorts for summer tourists, and as a sanitaria for
consumptives and invalids generally.

Although rejecting, without hesitation, as at va-
riance with physiology and pathology, these high
mountain stations as winter residences for invalids, I
am prepared to accept them with certain restrictions
as summer sanitaria, more especially for those who
cannot reach the shores of the northern seas, that is
for many of the inhabitants of central Europe. Many
years ago Dr. Lombard, of Geneva, in his well known
work entitled " Le Climat des Montagnes," pointed
out the value as health resorts in chronic disease in
general of the higher mountain elevations in Switzerland, and their utility in the restoration of general
health in chronic disease. That a residence of some
weeks, or months, in midsummer in the high mountain regions, be it the Engadine valley to which the
current of popular and professional feeling sets at
present, the Tyrol, the Andes or elsewhere, should be
productive of good to the general health of consumptives, and secondarily, to their disease, is, I should
say, highly consistent with our physiological knowledge. For such to be the case, however, the
elevation must be sufficient to avoid the great
summer heats, but not so great as to substitute a
winter temperature for the cool summer one sought ;
and the food and lodgings attainable should be good
and healthy.

From a pamphlet of Dr. Whitfield Hewlett, entitled "St. Moritz as a Health Resort, 1871," it appears that the village of St. Moritz, at present the favourite resort of tourists and invalids, is 6100 feet above the level of the sea. Dr. W. Hewett gives some tables of the temperatures of the years 1866-67-68, taken in the neighbourhood, which, he says, "give a very fair idea of the climate of St. Moritz." I have deduced from these tables the average maximum and minimum summer temperatures for the three years, which are as follows :

	June.	July.	August.	September.
Min.	40·9	42·2	43	35·7
Max.	61·3	62·4	62·3	59·1

These tables give a July and August mid-summer minimum temperature a fraction lower than the mid-winter minimum temperature at Mentone for January and February, the coldest months ; and a mid-summer maximum temperature only, respectively, ten and seven degrees higher than the Mentone mid-winter maximum. At Mentone, from my ten years' averages, the January min. is 42·8 ; max. 53 ; the February min. 43·5 ; max. 55·7. These two months are the most trying for invalids at Mentone ; those during which the greatest amount of prudence as to clothing, warming, lodgings, and out-door exercise is required, and during which we have most frequently to treat accidental attacks of inflammation of the mucous membranes of the air-passages.

I would draw attention, also, to the fact that the change of temperature between the night mini-

mum and the day maximum at St. Moritz is too
great for safety, amounting as it does to 20°.   At
Mentone in mid-winter the difference is generally not more than 10°, except in March, when it
reaches 15°.

Not that I think an even temperature day and
night indicated in the treatment of chronic chest
disease.   On the contrary, I think such an uniform
temperature one of the therapeutical mistakes of the
day, and decidedly unphysiological.   All animated
nature requires, for healthy vital action, a change of
temperature in the twenty-four hours.   The earth
and its inhabitants are turned to the sun twelve or
more hours, but not twenty-four.   Plants kept in
heated houses at the same temperature do not thrive
—the night temperature must be lower than the day
if they are to do well.   Human beings, ill or well,
obey the same nutritive laws as plants ; ill or well,
they require a lower temperature at night than in the
day.   But a few degrees answer this requirement.
Twenty is too much, especially in localities where the
houses are badly built and draughty, and where there
are no proper heating appliances.

Such being the case, I think some lower elevation
in the Engadine valley, or elsewhere, preferable
to St. Moritz.   But I must refer for further details
respecting this region of Switzerland to Dr. W.
Hewlett's pamphlet, and to a very interesting one on
the same subject lately published by Dr. C. B.
Williams.   No one is better able than Dr. Williams,

who has carefully surveyed the whole Engadine district, to guide the profession.

Whatever advantages the Engadine possesses as a healthy high mountain summer station, are also possessed by other high mountain stations in different parts of Switzerland. Moreover, prudence forbids consumptives, even if they aim at high mountain elevations, to mingle with the stream of mere pleasure tourists, which has set in violently to the Engadine district of late, and often renders it difficult to get good food and good accommodation— absolute necessities for consumptive invalids. The hotels in the Engadine region in summer are constantly full to overflowing with tourists, and invalids have to get shelter where they can.

To the invalid population of central and eastern Europe, with only moderate means, the high mountain-elevations of Switzerland and of the Tyrol are invaluable, as by their assistance they are able to escape from the extreme and pernicious heat of a continental summer. But I advise our own invalid countrymen and countrywomen to stay at home, or to return home from the southern sanitaria in spring, or early summer, if they can afford to do so, notwithstanding the allurements of Alpine travel. They are safer in summer in country or farm houses, in the pretty quiet watering-places of our north-eastern coast— Norfolk, Yorkshire, Northumberland—in Wales, in the Highlands of Scotland, and in the more favoured regions of Ireland—than on the Continent. The

Swiss or Tyrol Alps can only be reached in mid-summer by lengthened travel, through scorched-up Europe, and when there the invalid has to run the gauntlet of mountain inns and roads along with the pleasure tourists, who often jostle him aside, and take the best places.   I believe that I owe my own re-covery in a great measure to having pertinaciously returned each summer to cool healthy England, avoiding the very great temptation offered by the Tyrol and the Alps to one who, like myself, delights in foreign travel and in glorious mountain and lake scenery.

In the Appendix I have given an account of my recent explorations in western Switzerland, 1878, which will probably be useful to those who, from health or other motives, are obliged to pass or choose to pass the summer months in Switzerland.

TEMPERATE WINTER CLIMATES.

To find a temperate winter climate we must leave the British Isles, and descend south.   Those who are suffering from Phthisis or chronic disease of the respiratory organs, and can do so, thereby, no doubt, very greatly increase their chances of recovery.   For seven months in the year, from the middle of October to the middle of May, not only is the temperature too low in Great Britain, generally below 55° Fahr., but it is all but constantly moist.   Cold moisture arrests the action of the skin, throws extra work on the lungs, and is a fruitful cause of cold, influenza, bronchitis, pleurisy, and pneumonia. Moreover, these influences increase, aggravate the inflammatory complications, and sequelæ of Phthisis. Indeed, a severe feverish cold, such as nearly all experience at least once in England during the winter season, may soften in a few days a great amount of lung deposit, and create a large cavity. Thus, in the course of these few days, the consumptive patient may pass from the first to the third stage of the disease. Indeed, chronic bronchial affections, whether existing alone, or complicating Phthisis or asthma, are all but constantly aggravated by the atmospheric conditions that reign in our winter.

The confinement, also, to which persons suffering from these affections must be condemned during the many months of bad weather which characterise a British or north of Europe winter, saps at the very root of constitutional improvement. After even a few days' confinement to the house the appetite, digestive power, and nutrition flag, and thus a barrier is raised to that amelioration of the general health, which alone can arrest the progress of the disease.

After devoting nearly twenty successive years to the study of the winter climate of the south of Europe, principally of the Mediterranean basin, after much travelling and reflection, after a careful perusal of the writings of other authors, I have come to the following conclusion :—The most favourable and accessible climates for chronic disease of the respira-

tory organs, and especially for Phthisis, as also for
all diseases characterised by organic debility, are the
more sheltered portions of the undercliff of southern
Europe, that is, the coast ledge which forms the north
shore of the Mediterranean from Toulon to Spezzia,
and the eastern shores of Spain. The climate of
these regions is all but identical, as evidenced by
their vegetation, with variations, differences explained by varying protection from the north.

For the various climatic and meteorological data
on which this opinion is founded I must refer to my
work on Climate, in which they are fully developed.*
I will only here state, that the winter climate in
these regions is exactly the one which would theoretically respond to the requirements of Phthisis, as
I have described that disease. It is cool, sunny,
bracing, stimulating, and dry. During the invalid
season, which may be said to extend from the 1st of
November to the 1st of May, there are on the Riviera
seldom more than about thirty days' rain, and probably less in eastern Spain. Thus, out of the one
hundred and eighty-one days comprised in the six
months named, about one hundred and fifty are
generally days of brilliant sunshine—so dry, that the
hours from breakfast to dinner may usually be passed

* "Winter and Spring on the Shores and Islands of the Mediter-
ranean; or the Genoese Rivieras, Italy, Spain, Corfu, Greece, the
Archipelago, Constantinople, Corsica, Sicily, Sardinia, Malta, Algeria,
Tunis, Smyrna, Asia Minor, with Biarritz and Arcachon as Winter
Climates." Eight Chromo-lithographic Maps; Forty Engravings.
5th edition. 8vo; pp. 654. Messrs. Churchill.

with perfect safety by an invalid, lying on the rocks
on a cloak, in the sunshine. During more than half
the days of rain, also, it is only partial, and several
hours of sunshine are enjoyed.

In such a climate, if the rules which I have laid
down in my climate book, for the guidance of invalids, in this to them unknown region, are strictly
adhered to, there is an energetic stimulus given to
organic vitality, and if it is not too late, or altogether unattainable, the powers of the system are
effectually roused. The appetite and digestion improve, assimilation and nutrition become more natural,
the progress of disease is arrested, and nature at once
begins to repair existing mischief. In Phthisis, under
the influence of this constitutional improvement,
crude lung deposits are often absorbed and reduced
to their inorganic cretaceous elements, cavities cease
to secrete muco-pus, and then contract and cicatrise.
These are results which I witness every winter, in
many cases, although of course by no means in all, at
Mentone, certainly the most sheltered and favoured
spot of the whole Riviera. In some the disease is
too far advanced, or the organic taint or exhaustion
is too profound, for any stimulus—hygienic, climatic,
or medicinal—to favourably modify the constitution,
and thus to arrest the onward progress of the malady.
The rapidity only of its progress is modified, and it
terminates by death, as in the ordinary run of cases
in which all these means of treatment are not, or
cannot, be applied.

There are cities in the south of Europe, such as Naples, Rome, Pisa, Malaga, that have long enjoyed an exceptional reputation in the treatment of pulmonary Consumption. I believe that the climate of all these southern cities is very inferior to that of any part of the Riviera. I would except Malaga, which ought, however, to be avoided on account of its very filthy and very unhealthy state. They are all, without any exception whatever, very inferior to the western regions of the Riviera on hygienic and health grounds. They are all dirty, unhealthy, southern cities, with a very high rate of mortality from the diseases which produce the same results in the worst parts of our worst cities.

I maintain that consumptive people should reside in the country or in the suburbs of healthy towns, in order to secure the most favourable hygienic conditions. These favourable conditions are not obtainable by those who live in the centre of badly-drained unhealthy towns. On the Riviera, at Hyères, Cannes, Nice, Monaco, Mentone, St. Remo, the houses occupied by invalids are nearly all suburban, in the country, with the sea in front and mountains behind.

The great winter sanitarium for consumptive invalids up to the present day has been Madeira. With all its charms and all its advantages, it does not appear to me to offer the conditions which are indispensable to rouse exhausted vitality. It seems rather to satisfy the requirements of the bygone period of the professional mind, when pulmonary

Consumption was treated, all but entirely, by lowering or antiphlogistic means, as a mere species of inflammatory disease, than to satisfy present requirements. As I have repeatedly said, Phthisis is unquestionably a disease of debility, of anæmia, of organic exhaustion, and of defective nutrition. According to the former view, a moist, mild atmosphere, a kind of natural orchid house, would be just the place chosen. According to the latter, such a climate should rather be avoided, as calculated to depress vitality. The immunity from colds and inflammatory diseases of the respiratory organs which a warm, moist climate is calculated to afford, is purchased too dearly if it is gained by a loss of general tone, and by a diminution of appetite and nutritive power.

A recent writer on Madeira, Dr. Stone, of the Brompton Consumptive Hospital, says, in the pages of the *Lancet* (December 2nd, 1865) :—

" . . . . The first effect of the climate of Madeira is peculiarly soothing. It is not until some months have elapsed that the balmy influences of equable temperature, and the soft breathing of moist, warm sea-breezes become absolutely cloying, and tend to enervate both mind and body. Some temperaments resist the approach of this *dolce far niente* longer than others ; not a few, with well-meant efforts at resistance, pay by feverish attacks for unnecessary activity purposely indulged. . . . . All local varieties, however, are subordinate to the dominant character of the climate, which is warm, equable, and moist almost to saturation."

The slightest consideration must lead any physiologist to the conclusion, that although such a climate may be a very agreeable one, may be especially soothing to all who are suffering from chronic bronchial disease, idiopathic or symptomatic, it cannot rouse organic vitality as the more trying, cool, dry, sunny climate of the undercliff of the Riviera undoubtedly does.

At Mentone we have often, during November, this kind of weather, just like that of Madeira I am told by those who have long resided there, with warm south-westerly winds. The air is mild, moist, and balmy, and the weather perfectly delightful. Every one is in ecstasy at the unexpected prolongation of summer, every one enjoys the warm soft days, the lovely mild nights. But during the persistence of this beautiful weather the consumptives do not improve, but remain languid and ill. With them improvement commences when the wind veers round to the north, when the nights become cold, and the days cool and bracing, with a dry, crisp air, flooded with sunshine.

In the year 1859, impelled by the spirit of self-preservation, imbued with the views respecting the nature of Phthisis which I have propounded in this essay, and which I believe are the views of the more advanced members of the profession at home and abroad, I carefully analysed the claims of the various winter stations. I thought of Madeira long and seriously, but shrank from it on the above-mentioned grounds. Already well acquainted with the south of Europe as a traveller, I thither directed my steps as an invalid, and I believe that I found in the Genoese Riviera the regions that correspond with the medical ideas of the day. These, the modern views, are in advance of the medical ideas of former times, and in my opinion the expression of truth.

Since then, as stated in the last edition of my work on Climate (the 5th), I have found the same conditions on the eastern coast of Spain. This region also satisfies the requirements of consumptives, with reference to temperature and climate generally. But as yet there are so few material resources, there has been so little preparation for invalids, that it is a mistake for them to settle there. I must also refer my readers to the work mentioned for details respecting Corsica, Sicily, and Algeria—all good climates, but inferior to the Riviera and the eastern shores of Spain.

I would merely state, briefly, that the north shore of the Mediterranean, and the east coast of Spain, are necessarily dry, cool, sunny, and bracing in winter, inasmuch as north, north-eastern, and north-western winds, or terrestrial winds, reign during the greater part of the winter. These same winds become moist when they have crossed the Mediterranean and reach the southern shores of that sea. Thus the

islands of Corsica and Sicily, and Algeria, must have
and have a mild but moist winter atmosphere.

Dr. Stone, in the article I have quoted, thinks
there is a fashion in these things; that formerly
Madeira was the fashion, and that now the Riviera
and Mentone are becoming the fashion. I believe
that he is mistaken. Madeira was supported by the
medical profession as long as it thought that such a
climate answered its requirements, which were then
moisture and warmth. Now that it does so no longer,
that other views prevail, that a cool, bracing, toni-
fying climate is demanded, Madeira, the West Indies,
Nassau, Bermuda, fall in professional estimation, and
the Riviera and Eastern Spain will take their place.

I may lay claim to having been, during the last
twenty years, as a result of my own break-down in
health, once more a medical pioneer, and that in a
branch of medicine different from the one which long
occupied my attention and thoughts. I have sought
and found localities suited to the present improved
state of the professional mind abroad and at home.
All that I have written would have fallen to the
ground had I not met with a ready response in the
tone of mind of my professional brethren, who, after
all, alone direct public opinion in these questions.

Although I give a decided preference, in the treatment of chronic chest diseases, to the mild, dry,
sunny winter climate of the more sheltered regions of
the Mediterranean, I do not mean for one moment to
assert that such diseases do not get well, under judi-

cious treatment, in the colder and moister climates
of the north of Europe. What I do assert is, that
the chances of recovery are very considerably increased
in all cases by avoiding a northern winter, and that
in some recovery takes place when it probably would
not take place in our own climate. As I have stated,
the fine weather and sunshine of the winter on the
Riviera, on the east coast of Spain, in Corsica, Sicily,
and Algeria, enable invalids to spend the greater
part of their time out of doors, to the great improvement of their general health. In our climate, all
must admit that the frequent, often constant, rain,
the fogs and mist, the low temperature, may render
it impossible for great invalids to go or to remain
out of doors for days, weeks, or even months. Again,
the dryness and mildness of the atmosphere and the
mildness of the temperature in the south often ex-
tinguish chronic catarrhal or bronchial affections, both
in the consumptive and in the aged, which would be
incurable at home.

Notwithstanding these drawbacks, however, many
persons suffering from chronic chest disease, from
pulmonary Phthisis, chronic bronchitis, pleurisy, or
pneumonia, may and do recover without leaving the
British Isles, provided they abandon town life, settle
in the country or in some of our sanitaria, and are
judiciously treated. Hastings, Ventnor, Bournemouth, and Torquay are full of such cases. Many of
the residents, tradespeople, gentry, and doctors in
these localities are cured invalids, who have settled

where they recovered their health, just as I have settled in winter at Mentone. Most fortunate it is that such should be the case, for not two in fifty of those who suffer from chronic chest disease, the most frequent of all fatal diseases, are in a social position which enables them to incur the expense and sacrifices of winter expatriation.

The same may be said of the more sheltered parts of the south-west coasts of Scotland and Ireland. Owing to their contact with the warm waters of the Atlantic Ocean the climate is much milder than in the north-eastern and central regions. Many sufferers from chronic chest disease, and from chronic disease generally, recover their health in these localities. Much, however, of the benefit apparently derived from the climate is due, no doubt, as in the south of Europe, to the rest, to the hygienic life, and to removal from the causes of disease, mental and bodily, which operate at home.

The cases that are the most likely to recover at home are, I should say, those in which the consumptive disease, the phthisical deposits, have appeared from the first as a lung infiltration, exudation, or formation, without catarrhal antecedents, in persons not constitutionally liable to catarrh, to colds. A mere sthenic constitutional treatment in the north is more likely to be successful with them than with those who are constantly suffering, owing to some constitutional predisposition, from bronchial attacks, at every change of temperature or of season. To the latter the dry sunny climates I have mentioned, which protect them from catarrh or influenza, and remove or limit these affections where they exist, offer an additional and most valuable curative element.

This remark applies, specially, to those advancing in life, often children of gouty or rheumatic parents, who show a special tendency to chronic bronchitis. Such disease in them is, occasionally, only the precursor of phthisical deposits. In these constitutions, even if pulmonary Phthisis does not declare itself, the bronchial affection may become permanent, lasting winter and summer, lead to asthma, heart and liver disease, and render life itself a burden. Our workhouses and asylums for the aged are full of such forms of disease. They usually every winter get worse, the liver and the heart and the kidneys become diseased, and they at last become the victims of a species of pseudo-Phthisis, without discoverable morbid deposits. Such sufferers may be quite cured, or at least escape the gradual aggravation of their infirmity, by passing the winters in a dry, cool, sunny climate, like that of the Riviera or of the east coast of Spain.

It is worthy of remark that the presence of the traces of past phthisical disease which I, and other observers, have so frequently found in the lungs of the aged who die without having left their country, is an additional proof that Phthisis is quite capable of spontaneous cure in northern regions, provided it has not entered the more advanced or progressive stage of its existence.

When phthisical patients are sent abroad for the winter their medical attendants should strenuously warn them of the dangers of sight-seeing and of travelling for mere pleasure, as also against the rest-lessness which often takes possession of consumptive persons, especially in the earlier stages of their disease.

Nothing is less conducive to the welfare of any one really ill than continued travelling for pleasure, especially in winter. All kinds of trials and hard-ships have to be encountered, which are immaterial to those who are well, but which often result in sickness and death to those who are ill. A con-sumptive invalid, wintering abroad, should travel by easy stages to the locality selected, should settle in a good, well-built house, in a sheltered, airy locality, should place himself under trustworthy medical guidance, and should never stir more than a few miles from home, in fine weather, until the end of the winter—that is, until the first week in April or May. By adopting this rational course every daily incident of life is under thorough control.

With many such sufferers, however, restlessness is a part of their disease, and it is very difficult to restrain them. This is more especially the case with our American brethren, who are so alive to and so interested in all the natural, artistic, and social blandishments of Europe, that they are nearly always in a kind of feverish state of anxiety to be up and moving.

Many American invalids, consumptive and others, come to Europe every spring, with the full intention of seeing the Rhine, Switzerland, and Tyrol during the summer, and Italy, or Italy and Spain, during the winter, and all whilst attending to their health. But it is impossible to combine the two. The Continent of Europe is very hot—a furnace in summer; Italy is very cold at times—an ice-house in winter. So, what with the heat of the summer, the cold of winter, and the fatigue and trials of travelling, they are often worse in health at the end of the year than at the beginning. They clearly require this warning before they leave the States. Once arrived in Europe, the warning does not reach them, generally speaking, or if it does is not heeded. A very sad case that occurred a few years ago will illustrate the truth and importance of these remarks.

One morning, in January, 1867, a well-known American gentleman, wealthy, intellectual, refined, a tall handsome man of thirty-nine, called on me at Mentone. He was thin, emaciated, and looked ill. He said he had recently settled at Mentone with his family for the winter months, and, at the request of his home medical attendant, had called on me to report himself. He had had some little "lung trouble, or lung fever," for the last year, which had bothered him a good deal, and his physicians had sent him to Europe in the spring for a year or two. He had been travelling all over Rhine-land and Switzerland during the summer and autumn, but had recently been driven from the latter country by the severity of the weather. He could not say that he was much better since he had arrived in Europe, but he was not worse, and did not consider that he wanted medical advice. Only, as he had not been examined for some time, he wished I would just put the stethoscope on his chest, and tell him how he was going on. I did so, expecting to find a little chronic bronchitis, when, to my surprise, I found under the right clavicle a phthisical cavern, large enough to contain an orange, with a considerable amount of unsoftened deposit in the upper regions of both lungs. I thought it my duty to tell him, kindly but authoritatively, that he was altogether mistaking his case, that he was decidedly consumptive, that he had mismanaged himself since he arrived in Europe, and that all his thoughts and care ought to be devoted to the arrest of disease and to the pro- longation of life. His reply was that I took much too gloomy a view of his case, that he was certain there could be nothing seriously amiss with him, but that he would do what I wished.

I saw no more of him, professionally, for a month, although constantly meeting him driving and walk- ing, like a sound man. After that interval he again appeared, saying, "Doctor, I have come to you because I have a little spitting of blood, and I have brought some to show you." With this he took

from under a paper a jar, in which there was at least
six ounces of pure blood! I immediately sent him
home to bed, and soon followed myself. Under
proper treatment the hæmorrhage was arrested, and
on the sixth day the sanguinolent expectoration had
stopped. On this day he handed me over my fees,
adding, "Thanks for your kind care of me, but I
think I can now manage myself, at least for the present, and I will come and see you when I get out."
Again an interval of six weeks took place without
my hearing anything of him, when, on the 16th
March, he came in to me and said, "I have called to
ask you to take care of my family whilst I am away.
I am getting tired of this place, and start to-morrow
for Algiers, intending to meet my people at Pau
about a month hence." I tried, but in vain, to dis-
suade him from the voyage he purposed to take. I
told him that from the 15th to the 30th of March, or
5th of April, during the equinoctial gales, the Mediterranean was often, indeed usually, ravaged with
storms—a mere seething caldron ; that he might
have to take refuge on the coast of Spain, be four or
five days at sea instead of forty-eight hours, and
again suffer from hæmorrhage. He laughed at my
advice, said he was a good sailor, and feared nothing.
Clearly the mere lung-trouble view of his case was
uppermost in his mind, and he did not believe or
heed me.

Six weeks later, towards the end of April, I was
at Montpelier, on the platform waiting for the train

to Toulouse, when who should appear but my former
patient, still thinner and more emaciated than when
we last parted.   On my accosting him he smiled
feebly, and said : " Would, Dr. Bennet, that I had
taken your advice !   All you prophesied came to
pass.   We had a severe storm after leaving Marseilles, were obliged to take refuge on the Spanish
coast, and were five days out before reaching Algiers.
I was attacked with hæmorrhage on board, had to be
carried out to the hotel, remained there ten days,
and then left, fearing I should die away from my
family.   We again had a storm with the same results,
and I arrived half dead at Marseilles, where I have
been in bed ever since."

We each pursued our journey.   Two mornings
afterwards I was at the Toulouse station, seated in
the Bordeaux express then about to start, when this
gentleman's courier saw me and rushed to the door.
" My poor master," he exclaimed, " is dying.   He
was taken with hæmorrhage last night in the train,
has had two doctors with him all night, and is now
passing away."   I would have gone to his assistance
even then, but the train started before I could leave
it.   I heard at Bordeaux that he died that day, and
thus was a valuable life brought to a premature close.
I do not now know whether his ignorance of his own
condition was the result of his physicians having,
in mistaken kindness, veiled the truth from him, or
whether he resolutely shut his eyes to it, with them
as well as with me.   His year in Europe, however,
was most certainly entirely lost, for it was consumed

in the vain endeavour to combine pleasure and health, or sooner to make the latter bend to the former. The word lung-trouble, which I not unfrequently hear from my American patients, always brings to my mind the handsome intellectual face of this poor gentleman.

The restlessness which I am describing in those who go to the south of Europe in winter is often the result of disappointment as to the climate of the Mediterranean basin. They expect to find mid-summer instead of a bracing autumn, and move from place to place in the vain hope of finding mid-summer weather, which in reality does not exist, in mid-winter, out of the tropics. The mountains of Algeria are concealed in deep snow in winter, and the more northern regions even of the Desert of Sahara are often covered with hoar frost in January, at dawn of day. (See my work on Climate, p. 520.) Those therefore who want August days and nights, and whose case requires them, must go to the tropics or cross the line and visit the Cape of Good Hope, Australia, or New Zealand, where they will really find summer in January.

In some cases of Phthisis a long sea voyage is decidedly beneficial, and a voyage to the Antipodes may be recommended as a means of treatment even preferable to a winter in the south of Europe.

Thus young men without ties, in the early stages of disease, not unfrequently find such a voyage, in a large well-appointed vessel, a good mode of escaping the winter of Northern Europe. Indeed, I some-

times recommend this plan to young men on other
than health grounds, to keep them out of mischief
when I fear their discretion. The Australian summer, however, is so very hot that great care is required. Tasmania or New Zealand is probably a
better, because a cooler, climate to reach at the end
of the sea voyage.

Dr. Lombard, of Geneva, has recently published
(Paris, 1877) the two first volumes of an interesting
and exhaustive work on climatology (" Traité de Climatologie Médicale"), to which I would refer my
readers. They will find in this work many facts
confirming the views enunciated in this chapter
respecting the deleterious influence of extreme heat
and cold in all latitudes, and at all elevations, on
human life.

Many of the American consumptives whom I see
in the winter at Mentone, have an ardent and very
natural desire to go home for the summer, even when
they intend to spend a second winter in Europe.
This is more especially the case if they happen to be
good sailors and not to care about the sea-passage.
We are always, however, at a loss to find a cool
summer residence, in the States, where the minimum
in July and August would be between 50° and
60° Fah., and the maximum between 60° and 70°.
I wish, therefore, my American colleagues would
try to find out some such locality in their mountain ranges, at an accessible distance from New
York.

CHAPTER IV.

THE CLIMATE OF MOUNTAINS.

ATMOSPHERIC PRESSURE—ITS EFFECTS ON RESPIRA-
TION—EXTREMES OF HEAT AND COLD—RAIN AND
FOG—PHTHISICAL MORTALITY AT HIGH SWISS
ELEVATIONS—IN NORTHERN EUROPE.

IN the preceding chapter I have stated that I agree
with the profession in general in attributing great
value to mountain air, to the mountain sanitaria of
central Europe, of Switzerland, of the Tyrol, in the
treatment of chronic disease. But I have also added
that I consider there is a tendency at present to
exaggerate the beneficial influence of mountain air,
especially in the treatment of chronic pulmonary
disease. The question is so important an one that I
purpose devoting a special chapter to it.

Wishing to again test the results obtained in
former travel, and being free from professional duties
this summer (1878), I spent the greater part of June
and July in Western Switzerland, visiting and carefully studying above twenty of the best known and
most frequented mountain sanitaria. During this
exploration I did my best, by observation and reflection, to come to some definite conclusion on the
subject. To do so, however, more is required than
can be done by personal observation. The experience

and testimony of cautious observers must be carefully
analysed and weighed, and the physiology of mountain climates in the healthy must be ascertained.
This I have tried to do, and I present my readers
with the results.

I admit that mountain air offers great advantages,
such as purity of atmosphere, freedom from noxious
and malarious exhalations, a cool night temperature
(inestimable for those who leave hot plains), purity
of water, sparse population, and out-door life. All
these conditions, however, can be obtained in temperate climates, without ascending mountains.
Moreover, past travelling experience has long ago
taught me that there are peculiar disadvantages
connected with all mountain climates in European
latitudes which partly neutralise the advantages,
such as sudden and unavoidable changes of weather,
sudden outbursts of fog, rain, and snow, even in mid-
summer, and, in valleys, diminished sunlight.

It is impossible to define and to describe mountain climates with reference to elevation above the
sea alone. The only element which is nearly the
same in all regions, at a given elevation, is atmospheric pressure, as indicated by the barometer. All
others, temperature, dryness, moisture, atmospheric
calmness or motion, vary in every region examined
according to latitude, to exposure, north, south, east,
or west, to the nature and configuration of the soil,
to the proximity of glaciers, rivers, lakes, to habitual
air currents, such as those which ascend valleys in

warm weather, to liability to cloud, to fog, to rain. Thus generalisation becomes difficult, if not impossible, and each region and country has to be studied by itself. The geographical fact that the perpetual snow line between the tropics in America varies from 13,000 to 17,000, is an apt illustration of this truth.

Undue importance is generally attached to the diminution of atmospheric pressure, and to the consequent rarefaction of the atmosphere in mountain elevations, by many who have written on the subject. It appears proved that less air entering the lungs at each inspiration the respiration becomes quickened, that is, more inspirations are taken in a given time, although not fuller ones. M. Coindet, a French physician, published in the *Gazette Hebd.*, 1863-4, several "Medical Letters" on Mexico, in which he gives an account of a series of interesting experiments which he made to elucidate this point. Thus he found on examining 250 Frenchmen at the sea level the number of the respirations in a minute was 19·36, whereas on 250 Mexicans in the Mexican plain of Anahuac, about 7500 feet above the sea, the mean was 20·297. He found the same difference in the pulsations of the heart. In the French the mean was 76·216 in a minute, in the Mexicans it was 80·24. His experiments show that the relation between inspiration and pulsation is normally as one to four, and this relation is not disturbed by mountain elevation. It is the same in both series of experiments.

On the other hand, mensuration of the thorax gave a rather unexpected result. In the French the mean was centim. 92·450, in the Mexicans 89·048. Thus the increased frequency of respiration does not appear to have been attended with increased ampliation of the chest at each inspiration, with increased muscular contraction. Had it been so the thorax would have preserved the evidence of constant increased ampliation, from freer and fuller respiration.

These experiments prove that diminished atmospheric pressure from rarefaction of the atmosphere is simply met, physiologically, by more frequent inspiration, and not by a more complete and more thorough inspiration as is constantly asserted. A mere physiological equilibrium is established.

It is owing to these facts that both man and animals adapt themselves, physiologically, to any elevation below the region of eternal snow. To be free from physiological disturbance, however, the change from a low elevation to a higher one, from the seaside atmospheric pressure to that of high mountain elevations, must take place gradually. In such a case the equilibrium is easily and thoroughly established.

The physiological influence of mountain air in elevated regions has often been erroneously interpreted from its having been studied on foot travellers. When they ascend mountains, not only is the change from one region to another rapidly effected, but there is also to be taken into consideration the fatigue, the great expenditure of force, and the physical disturbance and exhaustion that follow.

The aëronaut sitting quietly in the car of his balloon may be rapidly transported to a very great height without suffering. Biot and Guy-Lussac on the 29th Fructidor, 1804, ascended 7016 metres (23,000 feet) without suffering, except from cold and slight difficulty of breathing. MM. Barral and Bixio ascended from Paris on the 27th July, 1850, to nearly the same altitude (6755 metres), and felt cold, but no difficulty of breathing or pain in the ears. Mr. Glaisher, in September, 1862, reached the greatest elevation yet obtained (10,000 metres, seven and a half miles), and he and his companion all but died, but then his last thermometrical notation was 38·5 Cent. below freezing.

According to M. Gavarnet, one of the principal, perhaps the principal, cause of the dyspnœa, of the cephalalgia, and of the vertigo so often complained of by those who ascend mountains, is the overloading of the blood by carbonic acid. Were it the result of mere rarefaction of the air, aëronauts, sitting quietly in their car, would experience these symptoms pretty nearly to the same extent.

It is now universally admitted that organic combustion, both in the lungs and in the capillaries, is attended not only with the generation of heat, but also with that of force. In any great and continued muscular exertion, such as that of raising the body

many thousand feet up a mountain, there is a great
call upon the system for the generation of force.
Organic combustion is accelerated, more carbonic
acid is generated than the lungs can get rid of, the
blood becomes overloaded, and the symptoms enumerated—cephalalgia, dyspnœa, sonitus aurium—
appear.

In Mexico and in South America there are many
populous towns situated from 6000 to 10,000 feet
above the sea, where life goes on just the same as at
the sea level, without the rarefaction of the air producing any appreciable change in the respiratory
powers or in the health of their inhabitants. Mexico,
a city of 230,000 inhabitants, is nearly 7500 feet
above the sea. Humboldt, speaking of Quito, in the
Andes, 9492 feet above the sea, with a population of
76,000, says that he passed several months on the
plain in which it is situated without feeling that he
was living in so different an altitude. He adds that
seeing life go on just as in lowland plains and cities
—flocks tended, harvests reaped, physical labour
undergone—it was difficult to realise the situation of
this region, suspended nearly ten thousand feet in
the air.

M. Boussingault, writing to Humboldt, says that
it is impossible to deny that man is capable of physical adaptation to the atmosphere of the highest
mountains, after seeing the agility and power of the
torreadors at the bull-fights at Quito, after witnessing the activity and *mouvement* of towns such as

Bogota, Micuipampa, Potosi, situated at an elevation
of from 8000 to 12,000 feet. "He saw," he adds,
"young and delicate women dance all night in
localities nearly as high as Mont Blanc, where
Saussure had scarcely the power to consult his instruments, and where his vigorous guides all but
fainted with the exertion of making a hole in the
snow. Again, V. Jacquemont, the French traveller,
camped with his companions on the Thibet side of
the Himalaya at an elevation of from 16,000 to
20,000 feet, without suffering in any way. (See
Article, "Altitude," in the "Dict. Encyclo. des
Sciences Médicales.")

Such being the case, respiration being identically
the same in the rarefied air of mountain elevations
as in the lowlands, only quicker, and the chemical
composition of the air being the same, we must look
to other reasons for the acknowledged healthiness of
high mountain regions. These reasons are patent
to the most superficial observer.

In tropical regions all the unhealthy exhalations,
all the pernicious influences of the lowlands, are
avoided. At Vera Cruz, on the sea-board, yellow
fever and malarious fevers are rife, and very fatal.
On the Mexican plains, 7500 feet above the sea, the
former disappears, and the latter loses its deadly
character. Even in temperate European climates,
the plains become hot and unfavourable to health
in summer, whilst the high mountain sanitaria
remain cool and pleasant. They afford a valuable

and healthy resource to the inhabitants of the plains of central Europe, as our sea-shores do to those who live within reach of them.

To the latter, however, to those who can reach the shores of the sea in temperate climates, such as that of the British Isles, of northern France, of Holland, there is, in all probability, no real superiority in mountain air. The atmosphere is just as pure on the sea-shore in temperate regions as on the highest mountain elevations, and the necessary amount of oxygen is afforded to respiration in both regions, once physiological equilibrium has been established.

Indeed, the summer of mountain regions may be assimilated to healthy country or seaside localities in latitudes in which the mean summer temperature is the same. It is only with such lowland latitudes that any given mountain elevation can be compared as to climate, and as to its influence on health.

There are various causes that combine to give moderately elevated mountain regions, corresponding to temperate latitudes, a healthy climate favourable to life. The population is sparse, disseminated over wide regions, and their occupations are nearly always agricultural, keeping them in the open air. Moreover, the children, if at all delicate, die in winter in great numbers, my Swiss medical informants tell me, owing to the severity of the climate, leaving only the strong ones to grow up. This observation applies, of course, only to those who inhabit the higher moun-

tains all the year round. In European mountains
there is no resident population above 6000 feet, except
the monks of St. Bernard.

It is worthy of remark that some observers, whose
opportunities of observation have been very great,
such as M. Jourdanet, who practised many years in
Mexico, 19° 20′ N. lat., maintain that even in that
tropical region an elevation of 7000 feet and above
is detrimental to health and life, bringing on anæmia,
which he calls *Barometrical anoxyæmia*, the result of
deficient oxygenation of the blood. These views are
not, however, corroborated by M. Coindet, whose
experiments were made in the same regions, the
Mexican plains of Anahuac. M. Lombard asserts that
plethora and the inflammatory diathesis characterise
the pathology of medium alpine climates, but recog-
nises anæmia in the higher regions from this cause.

The influence of medium mountain elevations on
health in Europe, in the Alps, Pyrenees, and elsewhere
is clearly not altogether good, inasmuch as in many
regions it induces idiocy (cretinism) and goitre.
Their existence may, it is true, be partly explained
by other causes, poverty, deficient ventilation in close
houses and confined mountain valleys. I have myself,
however, seen these diseases very rife at an elevation of several thousand feet, as in the Val d'Aosta,
where the ascending and descending currents are
diurnal and brisk, thoroughly ventilating the wide
valley.

I have myself also noticed repeatedly that at great elevations, from five to six thousand feet in the Alps, the conditions, even in summer, are not favourable to health, or to those suffering from chest affections. Owing to the dryness and clearness of the atmosphere when the weather is fine, not only is the heat of the sun intense, but the atmosphere itself is warm. This year I was at the comfortable hotel at Murren, near Lauterbrunner, 5500 feet high, on the 15th July, on a lovely cloudless day, and found the thermometer in the shade 78° Fahrenheit at one o'clock, and at two 80°. The sun was fierce, broiling, whilst at and after sunset, at these altitudes, the thermometer falls rapidly, and the nights are cold. A short time before, on the 4th of July, there was a storm over this part of Switzerland, and in the morning there were three inches of snow on the Murren hotel verandah.

Whenever the weather is damp and rainy in the Alps at two, three, or more thousand feet, there are constantly clouds hanging on the mountain sides, or ascending the valleys ; these clouds are merely damp cold fog. Often there is snow at the higher elevations.

During my mountain excursions this summer in Switzerland, in the vicinity of the Lakes of Geneva, Lucerne, and Thunn, I repeatedly passed through these cold fog clouds, both in ascending or descending to mountain stations. They always produced in me pain in the chest at the time, followed by laryngeal or bronchial irritation. Indeed, there is no escaping them on European mountains at high elevations, even in July and August. Clouds or rain at and above the elevation of 3000 feet in mid-summer, often below, mean damp fogs, or even snow.

As already stated, I visited above twenty mountain sanitaria during these two months, at an elevation varying between 2500 and 5500 feet, generally sleeping at the place explored. At many there are handsome hotels, at all comfortable ones, even if rustic in character. I was amused to listen to the accounts given at each successive station of the peculiar advantages it possessed. If at 2000 or 2500 feet that was the proper elevation, all above were raw, bleak, windy, only fit for mountaineer visitors. When the elevation was between 3000 and 4000, if there were glaciers in view, Diablerets, Villard, Champéry, Engelberg, it was the right place, on account of the delightful cool healthy air that descended from the glaciers. If there were none in view, Morgins, Burgenstock, Sonnenberg, it was the right place, because sheltered from the dry, cutting cold ice and snow air from glaciers. If situated in a gorge densely clothed with timber, with waterfalls and cascades moistening the air, Giesbach, Weissenberg, it was the right place, because the air was so moist, all but saturated, not harsh, dry, bleak, cutting, like that higher up in regions near glaciers.

The fact is that at all these sanitaria a great amount of capital has been spent, and the doctors

I

and people connected with each individual place find very good scientific reasons for pronouncing their own locality and elevation the one that offers the best conditions for chronic disease in general, and for pulmonary disease in particular.

These one-sided views are often, if not generally, entertained and defended in perfect good faith, the concentration of thought on one locality warping the judgment. In science, however, we must learn to make allowance for local partiality, and try to be guided by purely scientific reasoning.

In the course of my summer's researches I met with a recently published and very interesting work, by Dr. Emil Muller, of Winterthur, entitled, "Der Verbreitung der Lungenschwindsucht in der Schweiz," 1876. It is principally statistical, and gives an elaborate account of the mortality from Phthisis in Switzerland generally. Amongst the tables at the end are a series giving the average mortality from Phthisis at different mountain elevations, calculated on data furnished by a considerable number of localities.

The average mortality at different elevations is given in three heads : 1. In persons whose occupations were purely industrial, principally watch and lace making, entailing confinement in workshops or at home ; 2. In persons whose occupations were partly industrial, partly agricultural ; 3. In persons whose occupations were purely agricultural. I here give a digest of Table XVI. :—

Average Mortality from Phthisis in the Mountain regions of Switzerland.

Elevation.	Feet.	Occupation.	
1. From 200 To 1600		Industrial . . .	10·2 per cent.
		Mixed	7·6 ,,
		Agricultural . . .	6 ,,
2. From 1600 To 2300		Industrial .	10·2 ,,
		Mixed . .	5·9 ,,
		Agricultural .	5·3 ,,
3. From 2300 To 3000		Industrial .	4·7 ,,
		Mixed . .	9·6 ,,
		Agricultural .	2·9 ,,
4. From 3000 To 3400		Industrial .	6·5 ,,
		Mixed . .	6·1 ,,
		Agricultural .	3·5 ,,
5. From 3400 To 4400		Industrial . .	9·8 ,,
		Mixed . . .	7·5 ,,
		Agricultural . .	5 ,,
6. From 4400 To 5000		Mixed .	7·7 ,,
7. Above 5000		Agricultural . . .	4 ,,

These tables are very instructive, and entirely negative the statements so often made of late years, that there is any special immunity from Phthisis in the mountain regions of Switzerland. They show that a certain proportion of the general population in the higher mountain regions die, as elsewhere, from Phthisis, the rate depending on occupations in life. Industrial pursuits, carried on indoors, in consequence, give a death-rate of 10·2, 10·2, 4·7, 6·5, 9·8, 7·7 per cent., according, no doubt, to the nature of the occupation. One of the highest factors, 9·8, is at an elevation of from 3400 to 4400 feet. At 4400 to 5000 feet, in mixed labour (partly workshop, partly agricultural), the death-rate from Phthisis is 7·7.

This is the lesson we learn in the plains, in all latitudes, in the north as well as in the south. The mortality of Phthisis rises in cities in all pursuits necessitating an indoor life—that is, with those with whom respiration is defective, with whom oxygen food is deficient. It rises according to the degree of closeness and to the nature of the concomitants. It falls in the country, in pursuits carried out in the open air, with perfect physiological respiration; would fall still more were not the night habitations generally close and unhealthy, even in the country.

Cold regions are no more exempt from Phthisis in Europe than temperate or mountainous ones. Thus, Phthisis is a common disease in Sweden, according to Dr. Lombard. In his recently published work, "Traité de Climatologie," vol. ii. page 110, he states that the mortality from Phthisis in the northern towns of Sweden is 147 in a thousand, in those of the centre 125, and in those of the south 131. The probable cause of this high mortality is the defective ventilation of the houses, universally warmed by stoves, and probably still closer in the north than in the south. Norway presents the same high mortality for Phthisis, 130 in the thousand, or one in 8, a very high rate.

In Denmark, the general mortality from Phthisis is 122 in the thousand, nearly the same as in London, where it is 121; lower than in Glasgow, where it is 158.

In Russia, from 1857 to 1859 the general mortality

from Phthisis was 164 in the thousand, in the four provinces of the north the proportion was lower, 119 ; but in the most northern, Archangel, the mortality was very high, 190 ; in the adjoining one west of Wologda, still higher, 204. It was comparatively low in that of Perim, 114, and still more so in Clonetz, 65. There are evidently causes at work to account for these differences, which do not appear on the surface.

Ventilation is always defective in rooms heated by stoves, respiration is incomplete. The products of retrograde changes are consequently imperfectly eliminated in the lungs ; and thus health is lowered, and Phthisis is engendered. Thence, no doubt, the frequency of the disease in central and northern Europe.

Extreme coldness of climate does not evidently preserve from Phthisis. Cold undoubtedly purifies the air, and may thus contribute to render respiration physiologically healthy, but this it only does with persons who are out of doors in the open air. In cold climates and regions, however, but a very small proportion of the twenty-four hours can be thus passed. Generally speaking, often for days and weeks together, life is passed in close unhealthy stove-heated habitations, under conditions most prejudicial to health. By living in winter at the top of a Swiss mountain 6000 feet high (St. Moritz) we merely reproduce the conditions of Archangel with a mortality from Phthisis of 190 in the thousand, or of Wologda with a mortality of 204 !

Iceland appears to be exceptionally free from Phthisis. The cause cannot be the coldness of the climate, inasmuch as in northern Russia, the climate of which is as cold or nearly as cold, we have seen the mortality reach 190 or even 204 in the thousand. Dr. Brehmer, who has written on the subject, attributes it to the constant use of fish oil as an article of diet. Acute inflammatory affection of the organs of respiration are not very frequent in Iceland, but a kind of epidemic bronchitis or influenza often appears, more especially in summer, which is very fatal. In 1862 it carried off 2·37 of the population, the principal victims being infants and old people (Lombard).

Phthisis appears also to be rare on the high plains of the Andes, in the cities of Mexico and Peru to which I have alluded. This undoubted fact is, no doubt, the origin of the present tendency to send consumptives to high elevations in Europe ; but the conditions are totally different. In these tropical regions of the earth, the perpetual snow line varies from 13,000 to 17,000 feet, instead of 8000 as in Switzerland. Thus an elevation of six thousand feet in Europe is a totally different factor to that of six or seven thousand in the Andes. I have been told by natives that at Quito (9872) the climate is warm in summer, mild and temperate in winter ; that the houses are large, with wide doors and windows, and that manufactories, and close, badly ventilated workshops, are all but unknown, the life and occupations being principally carried on out of doors. All these factors must be analysed and weighed before we can come to any conclusion as to the real cause of the infrequency of Phthisis in these elevated regions. I have some allies, medical and non-medical, at Quito, from whom I trust to obtain some information in the course of time, but that must be reserved for another edition of this work, should it ever reach one.

I believe I am warranted in saying that Swiss physicians generally do not send their patients suffering with chronic chest affections to the higher mountain elevations of their country. They keep them in mid-summer at elevations between 2000 and 4000 feet at the highest, such as Weissenberg (2940) and Gurnigel (3783). They fear the rawness and extreme dryness of the atmosphere at higher elevations in fine weather, and the constant fog, rain, and snow in bad weather.

They fear, also, the great variations of the thermometer between day and night, and the lowness of the temperature at the latter period of the twenty-four hours. They say that these conditions often produce acute and subacute affections of the aërial passages, pharyngitis, laryngitis, bronchitis.

CHAPTER V.

THE MEDICINAL TREATMENT OF PHTHISIS.

NO ANTIDOTE FOR PHTHISIS—THE MEDICINAL TREATMENT OF COMPLICATIONS—COD-LIVER OIL—IODINE—IRON—THE PREPARATIONS OF PHOSPHORUS—OPIATES—EXPECTORANTS.

I HAVE now reached the most difficult part of my subject, one that still affords great room for difference of opinion, even if the premises contained in the previous chapters are admitted. It would be vain to endeavour to reconcile the conflicting views which reign in the profession respecting the therapeutics of Phthisis, so I shall confine myself to a statement of the conclusions at which I have arrived from my own experience in practice.

As that experience has increased, I have gradually arrived at the conviction that there is no medicinal panacea for pulmonary Phthisis, tubercular or inflammatory, any more than for any other form of tuberculosis. There is no one remedy, in my opinion—no one drug that can act as an antidote to the morbid diathesis; neither cod-liver oil nor pancreatic emulsion, nor iodine, nor iron, nor the preparations of phosphorus, nor any other pharmaceutical agent.

Those who believe that there is such an antidote appear to me to ignore the very nature of these diseases, not to be aware that whether the morbid lung-deposits be tubercular or inflammatory, especially in their advanced and progressive state, their presence is merely the local evidence of exhausted vitality, of general vital decay—a mode of death manifesting itself as the result of worn-out organic power.

Such a condition is not to be remedied by physic, but mainly by physiology, with physic as an adjuvant, a handmaiden. It is only by removing all the causes that are depressing life, that are contrary to the healthy development of the functions of life, and by placing the sufferer in the most favourable hygienic conditions for the welfare of his organisation, that we can hope to arrest or cure such a disease. Here, again, horticulture has been of use to me. If a plant is failing because it is of a bad stock, or because it is placed in conditions of air, moisture, sun, shade, or soil, unfavourable to its habits and nature, it is not by adding this manure or that to the soil in which it grows, that it can be restored to health. All such efforts are vain. Its nature and habits must be studied, and then the conditions favourable to its healthy development in every respect must be adopted. Once this is done, a favourable change may take place, provided its vitality be not already too far depressed, or provided disease has not advanced too far to admit of recovery. At the same time, well-chosen manures, the addition of a necessary element deficient in the soil, may materially help the agriculturist or horticulturist.

So it is with physic in Phthisis. Although no mere drug can give new life to a decaying organisation, can arrest and cure a disease in itself a mere symptom of such decay, an enlightened use of medicinal agencies may do much to aid improved hygienic conditions, in rousing and restoring vitality, and in arresting the progress of disease. There are many stumbling-blocks in the path of consumptive invalids, many conditions of disordered functional activity, which render the most hygienic treatment nugatory, and which physic has the power to modify and remove. Such are disordered conditions of stomach, liver, and intestines, morbid states of innervation, cerebral and spinal, uterine, vesical, rectal complications, functional or local, all of which are more or less under the influence of medicine and surgery.

To meet these and other complications we have numerous and valuable medicinal agents at our call: mineral acids, alkalies, vegetable bitters, sedatives, narcotics, alteratives, astringents, all of which in turn do good service in the hands of the experienced physician. There are few stages or conditions of the disease in which such a practitioner does not find an important indication, something to do medicinally, by which nature, hygiene, and climate may be materially assisted. I am a firm believer in physic, and seldom or never leave my consumptive patients entirely to nature. I firmly believe that I can help them by the application of rational therapeutics, and try to do so. When myself a consumptive invalid, for many years I was always doing something in the way of medicinal treatment, and I have the decided conviction, right or wrong, that I much increased my chance of recovering by thus meeting the varying phases of my own case.

Having laid down on a broad basis the principles which should, in my opinion, regulate the treatment of pulmonary Phthisis, I have a few words to say on some of the therapeutical agents which stand highest in the professional mind, pre-eminent amongst which is cod-liver oil.

Professor Bennett of Edinburgh first introduced this agent to the profession in Great Britain, in a work written, *ex professo*, on the subject in 1841.* He had found it extensively given in Germany; and in this work communicated to his countrymen the experience of his German friends, as also his own. Some years later, Dr. C. B. Williams gave cod-liver oil the sanction of his experience. From that time its influence in favourably modifying nutrition, and in arresting the formation of phthisical deposits in the lungs, has been universally acknowledged; so that now it has become the great remedy for consumption, and that most deservedly. Some of our American

---

* " On the Oleum Jecoris Aselli." By Dr. John Hughes Bennett. Edinburgh, 1841. Same with appendix, 1848.

brethren state that, since its general use in Phthisis, the mortality from the disease has sensibly diminished, and that the general death-rate is lower.

The question naturally presents itself, if cod-liver oil undoubtedly exercises a beneficial and even curative influence on pulmonary Phthisis, why and how does it produce this effect? Chemical analysis of fish-oil does not give a clue, for the amount of iodine and bromine discovered is so infinitesimal that we can hardly admit that theirs is the potent influence; especially when we find that, administered alone in these or even in larger doses, the therapeutical effect is not produced. To discover the clue we must fall back upon physiology.

It is now generally admitted by physiologists that fatty substances, if not absolutely essential to digestion and nutrition, exercise a most beneficial influence over these processes. Indeed, nature appears to have placed fats within the reach of man all over the world, and to have implanted an instinctive craving for them in mankind. In northern climates, the natives consume largely fish-oils alone, or with their food; in temperate climates, butter and meat-fats take their place; while in subtropical regions vegetable oils, such as olive-oil, form an important element of the food. Even in the tropics there is the palm-oil, and gee, or butter, to satisfy the absolute want of fatty substances. From physiological requirements to those of the morbid condition of nutrition which constitutes Phthisis there is only a step, which observation has made. It has been long remarked that in these morbid states of the human economy a larger amount of fatty nutritive elements than is usually required, becomes a means of restoring nutrition to a more healthy state, and that these fatty elements become, positively, therapeutical agencies.

If this view of the action of cod-liver oil is the correct one, if in giving it we are merely ministering, in an exaggerated degree, for therapeutical purposes, to a natural health requirement, any fatty substance would have the same result. Within certain limits I believe that such is the case—cream, fat meat, vegetable oils, bacon, butter, all answer the physiological condition, and I invariably give them, if possible, when the patient's stomach cannot bear the fish-oil. But I also believe with Professor Bennett that the fish-oil is the best, is the easiest digested and assimilated, and is the fat to which the stomach gets the soonest reconciled, and which it can take the longest. I myself took an ounce and a half a day for five years without intermission, at last with pleasure, and always with benefit to the digestive processes. A medical friend, well known to the profession, who has like myself saved his life by the combined influence of hygiene, climate, and physic, could never take cod-liver oil; "but then," says he, "I took fabulous quantities of butter with my meals."

It is a known and admitted fact that the greater number of those who now recover from Phthisis are persons who have taken cod-liver oil. This fact certainly redounds to the credit of the remedy, but it must be remembered that those only can take it in whom the digestive organs are in a sound condition naturally, or in whom they have been restored to a sound condition by proper medical treatment. Women in whom uterine disease sympathetically produces nausea and sickness, those who are suffering from chronic dyspepsia, or from chronic liver or kidney disease, generally speaking, cannot take the fish-oil ; it nauseates them, makes them sick, and destroys their appetite, as often do all other fatty substances. Thus, the recovery of those who can, and do, take cod-liver oil may be not so much because they take it, as that their digestive system is sound, and that they can take and digest fat and plenty of good nourishing food besides. On the other hand, those who cannot take the oil, and die, may die not so much because they cannot take the oil remedy as because their digestive system is bad, and cannot be restored to a healthy state, so as to admit of the food-cure.

The undoubted improvement of the majority of the consumptive patients who can take cod-liver oil or other fats, in health, in strength, and in condition, has received additional and most valuable explanation and confirmation from some recent interesting physiological experiments. These experiments have been made during the last few years by Dr. E. Smith, the Rev. Professor Haughton, Dr. Frankland, Professors Fick and Wislicenus of Zurich, and others, with a view to arrive at a clearer notion than we had before respecting the origin of the power given out or spent by animated beings. They have been carried on under the influence of modern views respecting the correlation of physical forces, and the doctrine of the conservation of force and of the equivalency of heat and mechanical force. The generally received physiological idea of nutrition has long been that nitrogenous or albuminous food, by assimilation, is transformed into muscle and force ; whereas carbonaceous, fatty, amylaceous food, is burnt, and generates animal heat. The experimentalists whom I have quoted appear to have satisfactorily established : that the production of the muscular power spent by animals and man is not so much to be attributed to the assimilation of nitrogenous food as to the slow combustion of carbonaceous food. According to this theory, the formation of animal heat by the combustion of carbon is attended with the development of "force," of which the muscles are only the instruments, not the producers.

This view may be familiarly explained by the steam-engine. The latter, in burning coal, does not only produce heat, but power, the power that drags the train along. In a more obscure, but equally evident manner, the slow combustion of food in the processes of nutrition is attended with the development not only of heat, but of power, force. If the above views are correct it would follow, singular as the statement appears, that more power or strength is to be got out of fat than out of meat or muscular tissue, and this really seems to be the case. Tyrolese chamois hunters find that they can endure greater fatigue on beef-fat than on the same weight of lean meat. Accordingly, when about to absent themselves for several days in the mountains, they take beef-fat with them instead of meat. (See *Intellectual Observer*, July, 1866.)

Thus is explained the craving of mankind for fatty food, and for carbonaceous food generally. Thus is illustrated the generally acknowledged physiological principle, that man is omnivorous, and is also explained the strength of the rice-eating Hindoo, and of the potato-eating Irishman. A rational dietary is evidently the one in which nitrogenous and carbonaceous food are mingled in due proportion.

Lastly, we may safely conclude that fats are not "bilious," bile producers, as popularly believed, but that the inability to digest them is merely an evidence of defective, or of weak and easily disturbed digestive powers. The great majority of those whose digestive system is in good order digest fats with the greatest ease, and that in large quantities. The dislike so often shown to fat by persons in good health is often merely a result of education—of mothers most foolishly and erroneously picking out the fat from their children's food in early life "as unwholesome and bilious."

Iodine has a great reputation in the treatment of other forms of tubercular disease, and especially of scrofula, which may be said to be a form of the same disease. I presume this reputation is a deserved

one; but, as iodine is always administered con-
jointly with a generous dietary, and with persevering hygienic treatment, it is very difficult to form
an estimate as to its real value. In pulmonary
Phthisis it certainly does not appear to me to exercise much influence ; and as it is apt to disorder the
stomach and to interfere with the appetite, I now
seldom give it internally. I constantly, however,
use it externally, over the diseased regions of the
lungs, as a counter-irritant, and to promote absorption of adhesions.

Once it is admitted that the treatment of Phthisis
ought to be sthenic, invigorating—not antiphlogistic
or debilitating—iron and its preparations naturally
present themselves to the mind. I have often administered them, and I believe with benefit, in the
stage of convalescence or retrogression, when tubercle
is no longer deposited, but in process of absorption
or cretefaction, and when the period of debility and
lassitude supervenes. I have also given them during
the acute stage, but I think without beneficial result.
Indeed, at that stage they appear to me, like iodine,
often to disorder the stomach, and to interfere with
the appetite and digestion. When I observe this
under any medication I at once stop the remedy,
firmly believing that food is of more value than
physic, if the choice is between the two. It is a
remarkable fact that no physicians writing on
chalybeate or iron waters recommend them for
active Phthisis. Indeed, the opinion that they are

K

not only useless but dangerous appears to prevail all but universally at chalybeate springs.

The preparations of phosphorus were first formally introduced to the profession for the treatment of scrofula and tubercular disease in the year 1851 by Dr. William Beneke, then resident physician at the German Hospital, Dalston, now Professor of Medicine in the University of Marburg. Dr. Beneke, a learned and distinguished German pathologist, published in *The Lancet* of April 19, 1851, an interesting paper "On the Physiology and Pathology of the Phosphate and Oxalate of Lime, and their Relation to the Formation of Cells." In this essay he contends that phosphorus is essential to healthy nutrition, and especially to the formation of nitrogenous tissues. He advises the administration of the phosphate of lime in doses of from four to eight or twenty grains daily—1st, in scrofulous ulcerations; 2nd, in infantile atrophy, connected with rickets and diarrhœa; 3rd, in tuberculous disease, more especially of the lungs, in its earliest stage. After entering into many interesting details respecting the part taken by phosphorus in the formation of cells, and describing its very marked influence in the healing of scrofulous ulcers, Dr. Beneke says, p. 434 :—

"As most of these ulcers occurred in patients afflicted with scrofula, the question arose whether the phosphate of lime really cured the scrofulous dyscrasia, or only a part of it. With regard to this point, from many observations, the number of which has much increased since, it resulted that there exists an undoubted intimate relation between scrofula and want of phosphate of lime, but that we are not able to cure the dyscrasia by the mere use of the phosphate. The same is the case with tuberculosis, a disease which is well known to be intimately related to, if not identical with, scrofula. In both kinds of disease, however, we shall promote the cure in the most efficient manner by the administration of the phosphate of lime, and I cannot forbear recommending its use as much as possible. In the following pages I shall give the explanation of these facts, which I hope will at once prove that the effect of the phosphate ought to be such an one as I imagined and really found it to be. I have especially to mention, that the waste of tissues, or, in other words, the want of formation of cells, was apparently less in many cases of tuberculosis and scrofula which were treated with the phosphate of lime, besides other remedies, than in those that were treated without the phosphate ; that the cure of tuberculous ulcers of the intestines was evidently promoted and even effected by the administration of the phosphate, and this remedy proved most efficient in cases of incipient acute tuberculosis, and even those which are well known to manifest at their commencement nearly all the symptoms of typhus. I need not enter here more deeply upon the special effects of the administration of the phosphate in these cases, if the one general fact is kept in mind, that it increases

the formation of cells, or prevents the rapid and
fearful waste of tissues."

It will be seen by the above extracts that although
Dr. Benèke warmly advocates the preparations of
phosphorus in scrofulous and tubercular diseases in
general, and in pulmonary Phthisis, he does not
recommend them to the profession as a positive antidote, as a panacea to Phthisis.   With the discretion
and clear-sightedness which pertains to so eminent a
pathologist, he merely asserts that they are calculated to assist in cell formation, in remedying one
of the elements of the general "dyscrasia" which
precedes and accompanies these forms of disease.

Some years later, in 1857, Dr. J. Churchill read
before the French Academy the first memoir in
which he claimed for phosphorus and its preparations
the position of actual antidotes to pulmonary Phthisis
and to tubercular disease in general.   This claim he
endeavoured to substantiate in a work published in
French in Paris—" De la cause immédiate de la
Phthisie pulmonaire et de son Traitement par les
Hypophosphites," 2nd ed., 1864, and since translated
into English.   The subject has naturally much occupied my thoughts, and during the last twenty years
I have administered these preparations to a large
proportion of those whom I have attended.

Were I only to quote the successful cases that I
have had under my care, the cases in which the
lung disease has been arrested and even cured, I could
quote many instances of cure, myself included, which

have apparently taken place under the influence of the hypophosphites, as they were long and constantly administered. But, on the other hand, I have quite as many, perhaps more, cases of death to narrate in patients whose condition admitted of recovery from the extent of the disease, and who perseveringly took the hypophosphites from the beginning to the end.

Were the preparations of phosphorus given really an antidote to the disease, and the cause of the recovery in the first class of cases, they ought also to have cured many of the latter, for they were all placed under the same hygienic and social conditions. The scrutiny and comparison of these cases of success and non-success, however, have left in my mind the conviction that the different results obtained are to be explained by considerations of general pathology, by the type of the disease, the constitution of the patient, the conditions under which it was generated, and that the patients were not taking a remedy that had the power to control antecedents and conditions unfavourable to recovery. It is worthy also of remark, that I have always administered either Dr. Churchill's own preparations, or salts furnished by his own manufacturer, so that there is a certainty as to the genuineness of the drugs used.

Although not admitting that phosphorus and its preparations are an antidote to pulmonary Phthisis, for I have seen too many cases of failure to be able to admit it, I believe that they constitute a valuable medicine in asthenic disease, including pulmonary

Phthisis. Their administration, also, is quite rational physiologically, and I may say also agriculturally. Phosphate of lime is one of the principal elements of our economy. It forms the bones, and is found in our tissues, and especially in the brain and nervous tissues. It is sound physiology and pathology to give freely to the animal system as food, or as physic, the elements of which that system is composed. If it is right to do so in health, it is equally right in disease. In Phthisis observation shows that it is judicious to increase the usual amount of fat given in the system, and observation seems to show that it is equally right to increase the amount of phosphates. Phosphorus is only contained in limited amount in our food, although it exists in so large a proportion in our system. Its administration in a disease of debility may therefore be compared to manuring an exhausted field. If corn is grown several years in succession in the same soil, the crop at last fails for want of phosphate of lime, which is necessary to form the grain. Add bone-dust or phosphate of lime and the corn comes up vigorously, and the grain forms healthily and well. It is in this sense that I give the preparations of phosphorus, and that I took them for several years.

I wish I could add that Dr. Churchill, enlightened by his own experience and that of others, now took a more sober and rational view of the value of these preparations, but I am bound, in the cause of truth, to say that he does not. He appears, both from his

writings and practice, not to recognise the constant
powerlessness of the hypophosphites to arrest or to
cure pulmonary Phthisis. A case which occurred in
1867 well illustrates the, in my eyes, irrational confidence which he still feels in the efficacy of the
hypophosphites as an antidote. In this case there
was no reason why they should not have succeeded
had they possessed the power he attributes to them.

In June, 1867, an American gentleman, Mr. H.,
aged thirty-eight, a New York merchant, called on
me in London, sent by his New York physicians.
He had at the apex of the right lung a well-marked
tubercular or phthisical deposit, extending anteriorly
to the second rib, and posteriorly over the region of
the supra spinata fossa, with dulness on percussion
and dry crepitation over all these regions. The left
apex was in a suspicious state : slight diminution of
sonoreity, bronchial respiration, prolonged expiration,
but no morbid sounds. He was thin and rather
anæmic, digestion indifferent, muscular strength
diminished. There were some phthisical antecedents
in the family, but until the previous winter he had
been quite well, enjoying fair average health. He
was a sensible, rational, intelligent man, and placed
himself entirely under my care for a year, promising
to do whatever I wished, to be entirely guided by
me. He had with him his wife, an equally intelligent, reliable lady, and his means were such as to
enable him to do what he liked. Thus everything
was favourable for fighting the battle of life.

After seeing him in town for a short time, the weather becoming hot, I sent him to the west coast of Scotland, to Tarbet on Loch Lomond, and Arrochar on Loch Long. He remained there till the end of September, when he again appeared in my London consulting-room. I was not, however, satisfied with the result of the summer. He had not improved, although not worse. He then went to Fontainebleau, near Paris, for two or three weeks, and arrived at Mentone towards the end of October. There he settled down near me, and submitted himself entirely to my directions. Nothing was omitted during the winter that could possibly be done to arrest the progress of the disease, or to rouse organic vitality. All, however, was in vain; he got slowly but steadily worse. The morbid infiltration extended lower down in the right lung, gradually invading the upper region of the left, and softening took place, sparsely, in the right lung. When spring arrived he was in a state of extreme debility, all but unable to walk upstairs, emaciated, anæmic, blanched. It was clear to me that tuberculization was extending throughout both lungs, and that the case was utterly hopeless.

His wife saw that her husband was gradually losing ground, and asked me to tell her the exact truth. "He had," she said, "a father and mother at New York who tenderly loved him, and to whom he was much attached, as also a child, and if he was to die he wished to go home to die with them." I told her conscientiously what I thought : that all our efforts had been vain, that the case was hopeless, that she must resign herself to lose him, and that such being the state of things he had better go home at the end of May, see his physician in New York, and pass the summer in some cool place up the Hudson with his family. Should my fears and anticipations prove groundless, should he rally, he could still return in the autumn.

My advice was acted on, and the places were secured for New York by the Havre steamer. He had been taking the hypophosphites with me steadily, and his attention had been directed to Dr. Churchill's practice by the first edition of this work ; so he naturally asked me whether he should see Dr. Churchill on his way through Paris. To this I cordially assented, expecting he would merely endorse my opinion and advice. To my surprise Dr. Churchill took quite a different view of the case, thought he could even then arrest and cure disease in such a state as described, and kept him in or near Paris throughout the heat of the summer, in order that he might himself administer the hypophosphites. But he died, of course, the following November.

In a sthenic or strengthening treatment, such as I describe, in the curable stage of the disease, opiates can have but little place. What avails it to allay irritation, to quiet cough, and procure sleep, if thereby the desire for food is destroyed, as is usually the case when opiate cough medicines are given ?

Is it not better that the patient should have a moderate amount of distress and discomfort and eat, if eating is life and fasting death ? In the latter stages of disease, when all hope of recovery is gone, and it is merely a question of soothing the last stage of life, then opiates become an inestimable blessing in judicious hands. There are, however, other sedatives—prussic acid, hyoscyamus, belladonna, conium, chloroform, and especially the hydrate of chloral— from which much ease may be obtained in the earlier stage of the disease, when there is still hope of recovery.

As to expectorants I cannot say that I have much faith or reliance in them. If muco-pus is abundantly secreted, it is better away, and nature expels it by the natural and then easy process of coughing. When secretion diminishes, as it does if the disease improves, and the patient coughs spasmodically to get rid of a sticky tenacious secretion which causes tickling and irritation, I do not see what good is done by loosening it, as the term is, by squills, even if they have the power to loosen it, which I doubt. The real remedy is an effort of strong will on the part of the patient to repress coughing until the natural action of the bronchial villi has pushed the muco-pus into the larynx, whence it can easily be expelled. My attention was first drawn by Professor Bennett to this fact : that the dry irritating cough for which expectorants are generally ordered, is often merely the result of actual improvement, and is best met by emollients, and by moral control exercised by the patient.

Of course I do not allude to the distressing cough which characterises the laryngeal form of Phthisis. In these cases there is tubercular deposit in and below the laryngeal mucous membrane, subsequent ulcerations, and distressing and constant cough, caused by the irritation thus induced and kept up. Great relief is given by the occasional application locally of a solution of nitrate of silver or of some other agent; but this treatment, as all others, is only palliative, not curative. The arrest of disease or cure can only be looked for or obtained by general treatment.

The local inflammations of pulmonary tissue around softening deposits, the local pleurisies which are the result of these deposits reaching the surface of the lung, are no doubt relieved by counter-irritation— by painting the chest with caustic iodine, by croton-oil liniments, by small blisters, even by dry cupping; but I question whether much real benefit is obtained by issues. Indeed, I think the pain and annoyance they occasion often counterbalance all the good done. The inflammation can only be radically cured by the natural subsidence of the causes of internal irritation, which the counter-irritation of the issue does not in the least control.

A volume might be written on the medicinal treatment of Phthisis according to these views; but I purpose limiting myself to the above brief general exposition, leaving the profession to fill it up themselves.

CHAPTER VI.

THE RESULTS OF MODERN TREATMENT—PROGNOSIS.

THE TYPES OF PHTHISIS—HISTOLOGICAL THEORIES.

Acute Phthisis—General Phthisis—Chronic Phthisis—Phthisis among
the Rich—Phthisis with Complications—Scrofulous Phthisis—
Localized Phthisis—Phthisis of the Aged—Gouty Phthisis—
Phthisis among the Poor.

HAVING in the preceding chapters briefly discussed
the nature of pulmonary Consumption, and described
its treatment by hygiene, climate, and medicine, I am
desirous, in conclusion, to say a few words on the
results which are to be obtained by the employment
of these means.

As already stated, by combining the various
agencies described, and which constitute what may be
termed the sthenic treatment of Phthisis, many
patients may be and are saved ; but many still die,
and must ever die. The question that I now purpose investigating, as far as possible, is ; who are
those who may live, and who are those whom all the
resources of our art are unable to rescue from death ?
The answer to this question can only be approximated
by referring to the laws of general pathology. By
analysing the type of the disease, the circumstances
under which it was generated, its stage of develop-

ment when discovered and first treated, and its complications, we may often arrive at an approximative solution, or in other words form a reliable prognosis.

The consideration of the type, that is, of the mode or form, in which pulmonary Consumption presents itself to our observation, leads back to the nature and origin of the disease. If we follow Virchow the types we must admit are inflammation types : the catarrhal, the pneumonic, the scrofulous, the tubercular ; the latter being limited to acute miliary tuberculization. If we follow Professor Bennett the disease is one : a tubercular exudation in varied stages of development, from the grey or yellow miliary tubercle to the large cheesy or cretaceous masses. If the latter view is the true one, we must look to the laws of general pathology for the types of the disease.

I feel disposed on clinical grounds, as already stated, to adhere to Professor Bennett's doctrine, which agrees with and explains the disease as we see it at the bedside of the patient, whilst Virchow's certainly does not. The early stages of phthisical deposits at the apices of the lungs, before softening and secondary bronchitis have supervened, appear, in many cases, to have nothing whatever of an inflammatory character about them. There are the physical signs of a morbid deposit in the structure of the lungs on percussion and auscultation, generally with a lowered state of health, but there is nothing akin to catarrh, or pneumonia, or scrofula. Later, inflamma-

tory complications supervene, but they are clearly
secondary. The new histological inflammation theory
offers no clearer or more satisfactory explanation to
my mind of what we see clinically in every stage of
the disease than did the doctrines of Broussais.

Nor does the inflammation theory lead to a correct
practical view of the pathology and treatment of
the disease. Most of the patients whom I see in
the winter at Mentone, labouring under pulmonary
Consumption and imbued with these doctrines, medical and non-medical, take a most erroneous view of
their case. They seem to think it a trifling indispo-
sition, one that will give way readily to a few weeks'
or months' leisure and treatment, one that need give
them little or no uneasiness as to the future. There
is nothing very serious or terrible, they think, and
not unfrequently say, in a slight catarrhal pneumonia,
in a patch of inflammatory consolidation at the apex
of the lungs. Even a large phthisical cavity, which
places them in the third stage of Phthisis, is merely
a scrofulous abscess that will soon get well. Again,
with this idea of inflammation in their minds they
are disinclined to adopt the sthenic treatment recommended in the preceding chapters. They are
disposed to coddle themselves, to shut themselves up
day and night in close rooms, to fall back into a mild
Broussaism; in a word, in treatment, as in ideas,
they have a tendency to revert to the doctrines and
practice of thirty years ago. Thus, by these histological theories, the onward therapeutical progress of

the last twenty years is endangered. Holding these views, I am pleased to find that in the able article on Phthisis, by Professor Bennett, which will be found in Dr. Reynolds's "System of Medicine" (Sept. 1871), the nature and bearing of Virchow's doctrines are forcibly questioned, discussed, and con-troverted, on purely scientific and histological grounds.*

I think I cannot give my readers a better idea of the present state of the question than by reproduc-ing two or three of the Edinburgh Professor's more important paragraphs. These extracts, coupled with those from Dr. Southey's work, given in the first chapter, will place clearly before my readers the present state of this important question.

" With regard to its mode of production, tubercular matter is first separated from the blood-vessels as a fluid exudation, forming by its first coagulation a molecular blastema. The molecules of which it is composed then aggregate or melt into each other to produce the tubercular corpuscles. These, if com-pressed together and formed slowly, constitute the indurated dense granulations described by Bayle ; but if separated by soft molecular tissue, produce the more common yellow tubercles. The idea that these bodies are invariably the result of cell proliferation originates from the erroneous hypothesis maintained by Virchow and his followers—viz., that all morbid

---

* Dr. Reynolds's " System of Medicine," vol. iii., 1871, Art. Phthisis.

products are derived from cells. In their attempts
to maintain this view they have mistaken the
occasional enlargement and proliferation of fibre-cells
in areolar tissue, first described by Lebert as fibro-
plastic cells, for tubercular granules, which they
describe as the essential elements of the lesion. It is
not in the pleura or peritoneum, however, where such
fibrous growths are occasionally seen, that the real
manner in which tubercle is formed can be well observed, but in the lung, where the disease is most
common and best characterised. There, all observation demonstrates that it originates in a molecular
exudation, which in consequence of diminished vital
power, seldom passes beyond the nuclear stage of
growth. It is this low type of *hyslo-genesis* that com-
municates to the exudation those essential characters
which form the foundation of tubercular or phthisical
disease."—p. 540.

The first theory (the Professor's own, p. 551)
supposes an altered condition of the blood, originating in a perversion of nutrition. This perversion,
as we have seen, has been considered by some to be
owing to vitiated air, by others to imperfect assimilation of food, and by others to an hereditary taint.
It has also been shown experimentally that it may
be caused in the lower animals by inoculation of
various morbid matters. All these, and indeed other
causes, may originate or co-operate in diminishing
the vital power of the individual, and directly or
indirectly produce weakness, feeble digestion, and an

impoverished blood. It is when in this condition that any accidental irritation of the lungs, often inappreciable and undetectable, causes a limited congestion here and there in the pulmonary organs, which terminates in more or less exudation of the liquor sanguinis. This exudation coagulating causes the miliary and infiltrated forms of tubercle previously described, which partaking of the diminished vital power of the organism, instead of being transformed into the pus characteristic of a similar exudation in a healthy person, produces the small, irregular, and imperfect bodies called tubercle corpuscles. Instead of cells which are rapidly produced, broken down, and absorbed, as in pneumonia, we have numerous molecules and bodies resembling ill-formed nuclei. In short, we have a chronic exudation in which the vitality is so lowered that it tends to disintegration and to produce the lowest kind of organic forms—*i.e.*, molecular granules and nuclei."

" The second theory" (Virchow's) " is one which, instead of ascribing tubercle to an exudation from the blood of low vital power, regards it as the result of increased cell development and multiplication of the included nuclei. According to this view, tubercular matter is a new growth, which, when we consider that it sometimes reaches the size of an apple, as in the brain, would demand for its production increased rather than diminished nutrition. Notwithstanding the desire of those who support an exclusive cell theory to trace tubercle, as well as

L

every morbid product, to some cell transformation,
the most careful and repeated investigations of histologists have failed to do so. According to Virchow,
however, upon isolating the constituents of a tubercular mass, 'either very small cells provided with
one nucleus are obtained, and these are often so
small that the membrane closely invests the nucleus,
or larger cells with a manifold division of the nuclei,
so that from twelve to twenty-four or thirty are contained in one cell ; in which case, however, the nuclei
are always small and have a homogeneous and some-
what shining appearance.'* This description of small
nuclei in the interior of cells, and the appearances
figured as constituting the structure of tubercle,
have, so far as we are aware, never been confirmed
by any experienced histologist. Tubercle is so
common a morbid product that if such indeed were
its constitution it ought to be seen at once ; but our
most anxious and repeated efforts have failed to
discover it, nor does there exist a single preparation
anywhere capable of demonstrating it. Cells containing many nuclei are very rare, associated with
tubercle, and when they do occur are evidently
dependent on the occasional irritation of texture
which is produced around the morbid products—
they are a result and not a cause. As a matter of
fact, therefore, not to speak of the theoretical im-
probability of a disease originating in weakness

---

* Virchow, by Chance, p. 476 and fig. 140.

commencing with increased power of vital development in the pre-existing tissues of the organism,
this theory must be rejected."

" In support of this last theory it is further maintained by Virchow and his followers that the term
tubercle should be limited to the minute indurated
granulations which, as Lebert originally pointed out,
are the result of increased nuclear growth in the
fibrous tissues—what he denominated fibro-plastic
corpuscles. The larger so-called tubercular infiltrations of morbid anatomists and practical physicians
they regard as chronic, or as they call them cheesy
exudations. Dr. Burdon Sanderson proposes that
tubercle should be called an 'adenoid growth,' and
it may be granted that a mass of molecules and
tubercle corpuscles, such as we have described, in a
fibrous tissue, may present a vague resemblance to
one of Peyer's glands. But a slight consideration
must show that these distinctions are more verbal
than real. It is not the occasional, scattered, and
rare indurated granulation with which we are so
much concerned as the extensive chronic morbid
deposit. Transferring or limiting the term tubercle
to the accidental granules, and calling the general
and essential morbid product chronic inflammation,
or adenoid growth, constitutes no real advance in
pathology. What we have from the first maintained
is, that we have to do with a *tubercular exudation*
which differs from an inflammatory and cancerous
exudation in its low vital energy and diminished power

of transformation into cell forms ; and this is the essential element of Phthisis Pulmonalis. Two recent French admirers of Virchow's doctrines have proposed to separate ordinary Phthisis from granular tubercle of the lung under the name of Tubercular Pneumonia,* and Niemeyer suggests for the term Phthisis chronic Pneumonia."†

I have been greatly interested and gratified to find one of the most eminent and experienced American physicians of the age, Dr. Austin Flint of New York, questioning on clinical grounds, like myself, the soundness of the new histological and inflammatory doctrine of phthisis.   Dr. Austin Flint made an important communication respecting the recent histological theories of Phthisis to the Academy of Medicine, on October 20th, 1870.   Speaking of these doctrines, he says :—

"To discuss their merits might seem presumptuous in one who is not a practical histologist.   But the great majority of those who are expected either to adopt or to reject views founded on histological data are not and cannot be workers with the microscope. They cannot be expected to confirm or correct these data by their own observations.   They must be content to compare the testimony of different observers, and judge for themselves according to their confidence in the observers.   They may be competent, however,

---

* Bérard et Cornil sur la Phthisie. 1867.
† Niemeyer on Pulmonary Consumption ; Sydenham Society's Translation.

to exercise their judgment as to whether the data are
sufficient to warrant the views which are thereon
based. They may even be better qualified to do this
than those who draw deductions from original observations; for experience has shown abundantly that
microscopical observations are peculiarly exposed to
error from the bias of speculation or preconceived
convictions."

In the above thoughtful and wise sentences Dr.
Austin Flint authoritatively claims for clinical medicine and for clinical observers the weight they undoubtedly ought to have in the decision of so
important a question as that of the nature and
pathological meaning of Pulmonary Consumption.
As he says, we are not called upon to surrender our
clinical knowledge of the disease to new views founded
on a debateable interpretation of minute histological
phenomena, if we find that the new doctrines do not
agree with or explain the clinical facts we are daily
witnessing. I am warranted in saying "debateable,"
for, as we have seen, histologists equal in skill and
experience to Virchow and his followers, altogether
repudiate their views on purely histological grounds,
altogether apart from the clinical interpretation of
the disease as it presents itself to our observation.

Dr. Flint objects, like myself, to the retrograde
therapeutical tendency of the views in question. He
says: ". . . . In the recent treatises by Bérard and
Cornil, and by Niemeyer, blood-letting is advocated as
a measure important in certain cases of the common

form of pulmonary tuberculosis—namely, that known as infiltrated tubercle. In the first named of these two treatises, even venesection is recommended, and in the second local bleeding by cups or leeches. Bérard and Cornil advocate the use of the tartrate of antimony. They advise confinement within doors during the winter season. Niemeyer goes further, and advises confinement to the bed whenever febrile phenomena are present. These measures of treatment are based on the pathological view which considers infiltrated tubercle as a form of pneumonia; they are employed, in other words, as antiphlogistic measures. Bérard and Cornil state that the therapeutical indications are nearly the same as in frank pneumonia; and Neimeyer states, as a reason for giving to the affection the name 'chronic catarrhal pneumonia,' that thereby 'its prophylasis and therapeusis are promoted.' "

Nothing that I could add would place in a stronger light the great danger to therapeutical progress of these recent histological views. I have no hesitation in stating that every one of the therapeutical indications above given is utterly and entirely unfounded as tested by my clinical experience. As explained in the preceding chapters, my consumptive patients live in the open air, sleep with their windows more or less open, wash the entire body daily with cold water, live on the best food and wine they can get, take as much of it as they can digest, and I never have recourse to any depleting or lowering agency of any description whatever; and yet on this system Phthisis is a totally different disease in my observation to what it was in my youth. I should have been dead many years ago had I been thus treated, on antiphlogistic grounds, had I been treated as I and others treated consumptives twenty-five years ago, and so would a crowd of friends and patients who now surround me, cured, convalescent, or enjoying life as invalids, with their disease arrested.

Every year's additional experience increases my firm clinical conviction that in treating pulmonary Phthisis we have not to deal with a mere local inflammation disease, but with a constitutional diathesis—a condition of lowered general organic vitality, which manifests itself by a local pathological condition quite foreign to ordinary inflammation, whatever be the histology of the minute morbid processes. To cure pulmonary Consumption you must reach the constitution, rouse, vitalise the general organism, improve the digestive functions, and thereby nutrition. All local treatment is useless, or worse than useless ; all medicines given as a panacea, as an antidote, are vain, nugatory. No therapeutical measures directed to modify the co-existing bronchitis, arrest or ameliorate the actual disease, no counter-irritation or local depletion arrests or ameliorates the chronic pneumonic conditions existing around the softened lung deposits, although they may ease pain and delude the patient and attendant into the belief that good is done. Again, how different the symptoms and condition of

the consumptive patient when any acute or subacute
pneumonia, pleurisy, or bronchitis occurs accidentally,
in regions of the lung not occupied by softened
phthisical deposits.   We then have at once all the
usual symptoms of these diseases as they present
themselves in non-phthisical patients.   It appears to
me that, clinically, bronchitis and pneumonia on the
one hand, pulmonary Consumption on the other, are
totally different diseases from the first stage to the
last.   To pursue the parallel between these diseases
would be to extend the range of this essay beyond
what was intended, but I cannot refrain from
drawing attention to the different sites they
occupy.

The favourite locality of pneumonia, acute and
chronic, is at the base of the lung, and it is here that
real pneumonic vomices form, except in old age.
Then acute pneumonia often begins in the apices, as I
saw on a large scale at the Salpêtrière among its aged
inhabitants in 1840.   In pulmonary Consumption the
disease nearly always begins at the apices of the lungs.
This is so much the case that Dr. Bowditch, of Boston,
United States, a physician of great and deserved eminence as a thoracic as well as a general pathologist,
in five hundred of his own recorded cases of Phthisis
only found five in which the disease evidently commenced at the base of the lung.   I quote from
memory, as I have not the paper by me, but I believe
that is the proportion given in a very interesting
essay published a few years ago, on the primary

localisation in the lungs of the morbid deposit in the early stages of pulmonary Phthisis.

All diseases are greatly modified in their symptoms and progress, as also in the results of the treatment to which they are subjected, by the form or type which they assume from their first development, and none more so than pulmonary Consumption. On the above grounds I prefer to study these types on a clinical basis rather than on an obscure histological one ; that is, I prefer to refer the types of the disease to general pathology, which best explains them. At the bedside I have tried in vain to discriminate between catarrhal, pneumonic, scrofulous, fibroid Phthisis, whereas the laws of general pathology have proved a safe guide.

In the acute type Phthisis may run through all its stages in a few weeks or a few months. I have seen patients seized with a series of febrile symptoms, having all the appearance of typhoid fever, and die in four or five weeks. On a post-mortem examination the lungs have been found full of miliary tubercles. No treatment has or can have any influence whatever on the termination of such a case ; the patient is destined to die from the first day. It is to this form of the disease that Virchow's school gives, exclusively, the name of tuberculosis.

In other equally fatal cases of acute Phthisis, although the disease does not assume the form of a continued fever, and occupies several months instead of several weeks, in passing through its successive stages, there is no lull whatever, no interval of arrest.

The lung tissue is progressively invaded by exudations, which rapidly soften, so that both lungs soon become a mass of advanced disease, and the patient dies without its having experienced any remission. What can medical science, climate, or hygiene do in such cases? Most surely is such disease a mere mode of dying. Indeed, it is a question whether the most active and judicious treatment much retards the fatal issue.

Acute Phthisis, in either form, is much more frequently seen in youth than in middle or advanced age. The disease participates in the vigour and energy of the vital functions in early life. It may be considered the evidence of a profound and final decay of vital power; so profound that there is no effort whatever made by the economy to contend with the evil that attacks it. The cause of acute Phthisis must be sought for in the exaggeration of all the causes, hereditary and social, that produce the disease, and perhaps in their concentration in the same individual. Thus, I have repeatedly witnessed it in persons who, with the hereditary predisposition or taint, have been exposed to extremely unfavourable hygienic conditions—town life, bad and scanty food, contaminated atmosphere, and great sorrows and cares.

Next in gravity to acute Phthisis is the type of the disease in the chronic form, in which the phthisical deposits take place, not in isolated patches, but all over both lungs simultaneously, at the apex, in the centre, and at the base. If the patient dies from other disease in the early stage of this form of Phthisis, the

lung is found studded with crude deposits in its entire
extent. When they soften simultaneously, as they
may do, the secondary bronchitis is generally severe,
and the constitutional symptoms are very marked.

This form of the disease may be said to occupy a
medium position between acute and chronic Phthisis.
There are many degrees of intensity, but the more
the case recedes from phthisical deposits localised at
the apex—the favourable type for treatment—the
more serious is the prognosis.

When Phthisis assumes the chronic type, as is
generally the case, most fortunately, an unfavourable
form for treatment is that in which the disease shows
itself in the midst of very favourable hygienic and
social conditions.   If a poor sempstress, half starved,
made to work eighteen hours out of the twenty-four,
in a polluted atmosphere, living in a state of constant
mental depression, becomes consumptive, common
sense tells us that the disease may have manifested
itself from the action of removable causes.   If she
can be placed under more satisfactory hygienic and
social influences, she may therefore, and often does,
recover.   But if, on the contrary, the disease appears
in one who has been bred and nurtured in the lap of
luxury—who has known no hardship, no privation, no sorrow—of course the prognosis is more
unfavourable.

It must be more difficult to arrest the progress of
disease in such cases, for probably the cause is some
strong hereditary predisposition, some defect origi-

nating with the progenitors, or some defective condition of individual innate vitality.   It is in such cases, more especially, that everything should be done that is humanly feasible to arrest the disease, that no agency should be left untried that can possibly rouse the vitality of the patient.   It is in such cases that he or she should be at once removed from the social medium in which the malady has been generated, in the hope of counteracting some unknown, unrecognised, and yet powerful home antagonistic influence. A change of climate is of inestimable value with these patients; indeed, it may be the only chance of arrest or recovery.   Everything that is done, likewise, should be done from the very first; no time should be lost, for the foe is a most formidable one from the onset of the attack.

A class of cases still less amenable to curative treatment is that in which there are serious incurable complications present.   Phthisis not unfrequently comes on in persons advancing in age, between thirty and sixty, who have led a hard life, who have taken large quantities of stimulants, and have exhausted perhaps an originally good constitution by excesses of various kinds.   With them the stomach is generally out of order, the liver is often diseased, and sometimes the kidneys.   What can treatment do in such cases?   The disease may be considered a general break-up of the constitution, and the most judicious and persevering treatment seldom does more than retard the fatal termination.   Such is the form of

Phthisis which so often terminates the life of the diabetic.

Again, Phthisis may attack at puberty those who during childhood have suffered from scrofula. This is a grievous and serious complication, but by no means so unpromising as those just described. Tuberculosis, or tubercular exudation, affects different organs at different periods of life. In infancy and early childhood it more especially attacks the meninges and the mesentery. In childhood and until puberty it attacks, in preference, the glandular structures of the neck, the extremities of the long bones, and the spongious tissue of the bones in general, giving rise to the diseases of the articulations and bones which characterise scrofula. In early life, during my Paris career, I had charge for two years of a scrofulous ward of eighty young females, from fifteen to twenty years of age, in the Hospital of St. Louis. They had nearly all glandular swellings, with or without scrofulous disease of the bones, ankles, knees, elbows; a sad assemblage these poor girls were.

On several occasions I carefully examined the lungs of all my young patients, for I was publishing the clinical lectures of my master, the celebrated Dr. Lugol, and was much interested in everything connected with the pathology of scrofula. He wished to establish the connexion between scrofula and pulmonary Consumption, and I found the evidence of localised phthisical deposits, apparently tubercular, in

the lungs of many of these scrofulous girls. These
lung deposits were met with more especially among
the elder ones, and that often although the patient
presented little or no evidence of their presence. Dr.
Lugol told me that he had long found this to be the
case with his young scrofulous patients. When
death, through accidental disease, afforded an opportunity for post-mortem investigation, the development
of phthisical deposits in the lungs of scrofulous youths
was more frequently observed in those who were
arriving or had arrived at puberty than in those who
were younger. Such deposits often remain crude
and dormant for years, but when they assume a more
rapid development and soften, the previous existence
of scrofulous disease stamps the case as serious,
although not necessarily as fatal. In these patients
Phthisis seems to appear, in its progressive form, as
a species of climax to the antecedent tubercular or
scrofulous affections. Crude tubercles co-existing
with scrofulous disease in the young, in the latent
form, are, thus, often present without symptoms, and
they may be absorbed, and the patient may recover
without their presence having been even suspected.
Where is the inflammatory element in such cases ?

The more favourable type of Phthisis, that in which
rational treatment is the most likely to arrest the
progress of the disease, and even to effect a cure, is
that which may be termed accidental Phthisis, in a
chronic form, localised to the upper regions of the
lungs. In this type of the malady there is, probably,

no very decided hereditary taint, the patient is seldom born of very aged or very sickly parents, and does not present very serious complications in other organs, the evidence of a thorough and irremediable break-up of constitution. Again, the disease in these cases generally manifests itself under unfavourable hygienic conditions, under the influence of over-work, sedentary town life, or harass, care, and anxiety. Sometimes in the most apparently luxurious and easy-going life some of these influences may be at work, so that appearances must not always be trusted. The habits and general life of persons moving in the highest circles, and having within their grasp every comfort, may be unhygienic. Moreover, they may be a prey, like humbler mortals, to cruel cares, none the less felt for not being acknowledged or recognised. Their nights may be sleepless, their days without joy; disappointed affections, social ties, or ambition may deprave digestion, and pave the way to the inroads of disease. In all such cases we can reasonably hope that the old saying, " sublatâ causâ, attollitur effectus," may be verified. If we can remove *all* the causes that are depressing vitality, and the disease is not in too advanced a stage of its development, we may hope, firstly to arrest its progress, and secondly to effect a cure, and that at any period of life short of extreme old age.

Pulmonary Phthisis in old age—not so rare an affection as is generally supposed—appears to me an all but incurable form of the disease, a mere mode of dying. I saw a number of cases of this form of

Consumption in the year 1840, when in medical charge
of the infirmary of the Salpêtrière Hospital, Paris.
The infirmary is fed by a population of 3500 old infirm
women, between sixty-five and a hundred, all living
within the walls of this magnificent institution. The
disease assumes the form of chronic bronchitis, but is
characterised, besides the stethoscopic and percussion
symptoms, by a most unearthly degree of emaciation.
In no other disease have I seen patients live in such
a ghastly state of emaciation. They become at last
like living mummies—nothing, literally, but skin,
bone, and "concealed" organs.

There is a form of Phthisis of which I have seen a
good many examples although it is not generally
described. It may be termed gouty Phthisis, and the
prognosis is rather favourable than otherwise, especially in its early stage. Consumption and gout are
considered by many physicians to be antagonistic, but
experience has proved to me that such is not the case,
and the discrepancy between the theoretical and the
practical view admits, I think, of easy explanation.

Gout develops itself, primarily, in persons of healthy,
robust constitution, who live generously. Their
digestive system being good enables them to take and
assimilate a considerable amount of nitrogenised food,
and of stimulants, which appear to be the cause of
gout developing itself. These people—the primarily
gouty—do not become consumptive, for their vitality
is high and antagonistic to a disease of debility.

But when such robust gouty people, who have

themselves developed gout in their organization by a
luxurious existence, marry late in life, as they often
do, and have children, they do not generate healthy,
robust children like themselves. Their children are
often delicate, without being positively unhealthy ;
they have weak digestions, and suffer all their life
from what may be termed gouty dyspepsia. If their
organization is not much tried they get through life
very well, and may reach old age, even when suffering more or less from low forms of gout. If, on the
contrary, they are much tried, body or mind—if they
are placed for a continuance under unfavourable
hygienic conditions, they fall below par, are liable to
suffer from inflammatory affections of the aerial
passages, which become chronic from lowness of
general tone, and phthisical deposits in the lungs
may follow. As I have stated, I do not think this
form of Phthisis an unfavourable one for treatment,
for the constitution received from the parents is often
originally a good one, merely weak and tainted with
the low type of gout, and a tendency to catarrh.
There is often great latent vitality to work upon. My
own case is one of this kind.

An all-important element in estimating the probable result of treatment—in forming a prognosis, in
a word—is the extent of lung diseased when the
patient is fairly brought under rational treatment. I
often familiarly compare the lung attacked with
Phthisis to a large house on fire. The fire may begin
in the servants' rooms or garrets—that is, at the apex

of the lung, the most frequent original seat of deposit. If it can be put out before it has extended to the story below, but little inconvenience is afterwards experienced by the owner of the house. He can live very comfortably in it under ordinary circumstances, only feeling that there is " less room" than formerly, on extra occasions, such as visits from friends. If the story below, or the two stories below, are destroyed before the fire is put out, he feels more or less inconvenience in his " daily" life, but still he can get on. But when all the house is destroyed except one room, or the cellars, it becomes quite impossible for him to live in it by any amount of contrivance. Moreover, it is all but impossible to save even that one room, when the fire has reached this stage.

So it is with the lungs, which are not restored, renewed, when once destroyed ; for we do not renew our organs as lobsters are said to renew their lost claws. Once a portion of the lung is destroyed, it is destroyed for ever, and its functions must be carried on by the healthy remainder. The only limit to curability, however, is the fact of there remaining a sufficient portion of healthy lung to carry on the functions of hæmatosis once the progress of the disease is arrested. The amount of healthy lung tissue compatible with life evidently varies in different individuals according to their vitality. One lives long on a bit of healthy lung no larger than a small orange ; another dies with even less than that in a state of disease.

The first and all-important point therefore is to arrest the progress of the disease, as it also is to arrest the fire in the house. Unless that can be done, in the one and the other case, the entire tenement will be destroyed, more or less rapidly. The living in the damaged tenement afterwards is a matter of adaptation; and it is wonderful what either Nature or man can and will do to adapt themselves to altered circumstances. We must also bear in mind that the more the disease or the fire has progressed, when it is first discovered, the more difficult it always is to arrest it.

When by the combined influence of hygiene, climate, and medicine the progress of Phthisis has been arrested, crude deposits have been absorbed or reduced to their mineral constituents, and cavities have been entirely or all but entirely cicatrized, it must not be supposed that the patient is well and safe. The recovery generally, always indeed, takes place through improved nutrition. Often the convalescent consumptive patient is fat and rosy, and looks healthy and well, but these looks are deceptive, the result of a life passed under the most hygienic circumstances possible, in unnatural quiet and repose. At the bottom there is still, all but invariably, the tubercular cachexia, which reveals itself by a want of power, by lassitude, and even prostration, if the habits of invalidism are abandoned, and the sufferer once more quits the shores of the stream of life for the rapid current.

M 2

Consumptive convalescents should consider themselves invalids for years, in many cases for evermore, and it is only by doing so that they can hope really to regain a firm footing in life. They may aptly compare themselves to a railway truck, " warranted to carry six tons," which after having been smashed, and then mended, painted, and varnished, looks as good as new, but is not so. It may carry two or three, or even four tons safely, but no longer the original six, under penalty of a final catastrophe.

Those who cannot or will not thus consider themselves invalids, despite the outer appearance of health, often relapse, and then frequently die miserably, for they are seldom saved. I have seen many such instances. One or two winters passed in the south, and rational treatment, arrest the disease, and bring with the improvement delusive confidence. The patient either cannot or will not listen to advice, and goes out again to fight "the battle of life," but only to relapse, and to return to the south in a hopeless state.

What proves that even in those in whom the progress of pulmonary tuberculosis is arrested tubercular cachexia or defective vital power long remains, is the frequency with which cachectic disease of another type subsequently attacks other organs. Most winters I lose at Mentone one or more patients from Bright's disease—cases of arrested Consumption. In some cases Consumption has been arrested for five or ten years, in others less. They generally

die with all but complete lung quiescence ; serous
infiltration gradually rises until it reaches the lungs,
and extinguishes life, or fatal convulsions supervene
from blood-poisoning.

Some years ago, some of the Parisian friends and
companions of my younger days, now men of mature
experience and occupying the more prominent positions in the Parisian medical world, gave me a dinner
as I passed through Paris on my way south.   After
we had dined, my case was talked over, and one after
the other gave the results of his experience of the
treatment of Phthisis.   All believed in its curability ;
all could quote cases of arrest and cure in their practice ; but one and all stated that many of these cases
of arrested Consumption had subsequently died of
some other form of cachectic disease, and principally
from albuminuria.   These statements exercised a
great influence over me, and prevented my returning,
when better in health, to active work in London.

Thus, perseverance and energy are long required,
not only during the course of treatment but for
years after, if a thorough recovery is to be made, or
even if a prolongation of enjoyable life is to be
secured.   This is really one of the most trying
features of the disease, even when successfully
treated.   If we succeed in escaping death we must
accept invalidism for a long period, perhaps for the
remainder of our lives.   I would remark, however,
that this applies more to the middle-aged who
recover from Phthisis than to the young.   The latter

have such an amount of organic activity about them,
the characteristic of early life, that if they recover
completely, they may, with care, and by leading a
hygienic life, regain a firm hold on life.

To secure this result I often advise my young
male convalescent patients to abandon, if possible,
sedentary pursuits, and to turn their thoughts to
out-door occupations.   Our Australian and South
African colonies offer valuable fields for such persons.
Life in the bush, among cattle and trees, in a dry
climate like those I mention, is certainly more
favourable to the prolongation of existence in a
convalescent phthisical patient than a city counting-
house.   Had I myself been a younger man, I should
have adopted this course.   As it is, I have approached
as near to it as possible by becoming an "amateur
horticulturist."   To our American brethren and
patients the change from sedentary to out-door life
and pursuits is still easier than it is to us.

Foolish people have scarcely a chance of recovery
—they must perish.   They generally do everything
that is wrong and pernicious to please their own
passing whims and fancies, and often look upon the
friendly physician, who tries to rescue them from
death, as one to be deceived and deluded.   Sometimes they turn upon him and pursue him as an
enemy who has done them bitter wrong, because he
has tried to save their lives.   I repeat, such unfortunate people have scarcely a chance of recovery.   They
have neither the sense to follow the right course

when it is pointed out to them, or to grasp the hand of fellowship and sympathy when it is held out; nor will they sacrifice pleasure, money, or ambition to the pursuit of life. Indeed, I consider a weak, vacillating, peevish tone of mind, or an inordinate appreciation of, and clinging to, the enjoyments and possessions of life, to be as unfavourable an element of prognosis in pulmonary Consumption as any of those already discussed. Such mental conditions all but certainly preclude recovery, however favourable the case may otherwise be.

When I reflect on the convictions that have gradually gained ground in my mind respecting the treatment of Phthisis—convictions embodied in the preceding pages—I am often saddened by the thought: how are the poor to struggle successfully with such a disease? If rest from weary labours, if protection from atmospheric vicissitudes, if ample, nay a luxurious dietary, if expensive medicines—such as cod-liver oil—if change of climate, to escape winter cold and wet, are necessary, how can those who live by their daily labour—and even many above them in social rank—hope to escape from the grasp of this fell disease? Is not the battle itself, for them, all but a hopeless one?

To these questions I would answer, that although the struggle for life cannot, most assuredly, be made with the same chance of success by the poor as by those whose position enables them to do all that is calculated to arrest the disease, yet their case is by no means a hopeless one. The means of treatment that I have recommended—hygiene, climate, medicine—may be attended to at home, in our own country, in the midst of the duties and occupations of life, although in a minor degree. I have met with cases of arrested and cured Phthisis in persons who have never left England, and who have never given up their social pursuits, and so have other physicians. Some of the most satisfactory and conclusive cases given in Professor Bennett's valuable work "On Pulmonary Consumption" are cases of this description. It is, however, worthy of notice that many of Professor Bennett's more remarkable cases of cure, he tells me, have been medical men. Fully able to appreciate the nature of their disease and its danger, they have probably been more energetic, more persevering in their attention to treatment and to health rules.

To attain the arrest and cure of the disease, however, nothing should be neglected. Unhygienic, unhealthy occupations should be given up, all the rules of hygiene to which I have alluded should be scrupulously followed, out-door occupations substituted for in-door during the summer months, if possible, and, more especially, town should be abandoned for the country as a residence, whenever feasible or possible.

Cities exercise a mysterious attraction over the lower as well as the higher classes of mankind. It must be the feverish excitement of city life, the hope of greater social advancement—for the greater portion of the lower classes in cities live as hard or harder lives than they would if similarly engaged in the country. No doubt the vitiated air breathed in cities, in the close crowded workshops, and in the closer and still more crowded sleeping-rooms, gradually weakens constitutional powers, and constitutes one of the principal predisposing causes of Phthisis. The poor should return to their native villages, if by any means feasible, even if they have there to accept a lowlier position than that which they have attained. The younger members of the family, when attacked with Phthisis, should be sent to board or work with country relatives. The country air would do them more good than all the physic they can get from hospitals and dispensaries in town, and give them a better chance of recovery.

Indeed it has often struck me that the funds of our city charitable institutions would be better employed by boarding their consumptive patients in farm-houses and agricultural villages, than in maintaining them in the wards of a city hospital. Or the hospital itself might be placed on some healthy, pine-covered moor, like the Convalescent Hospital at Walton-on-Thames; and out-patients only seen in town. Here, however, a public danger awaits us—so difficult is it to do good without evil following. Young consumptive towns-people who emigrate to the country and there recover and take up their abode, generally marry, and thus inoculate an otherwise healthy country district, as it were, with scrofula and Consumption. I recollect reading a lecture of Sir William Jenner's in which he states having observed this very fact in distant country localities.

It must be well understood that I am now speaking only of *curative* treatment. If all hope of recovery has been abandoned, if the lungs are all but destroyed, and the disease cannot be arrested—if an asylum to die in is all that is required—then it is of but little avail to drive the poor patient into the country, away from home ties and home nursing. Then, when the last scene is at hand, any asylum will do to die in— the small home, with dear friends around, the city hospital, the workhouse infirmary.

The recent researches regarding nutrition, to which I have elsewhere alluded, are consolatory as regards the poor. As long as we believed that, in the scheme of nutrition, meat meant muscle and strength, fats and cereals heat only, the poor at home seemed to have but a slight chance of recovery in asthenic diseases, diseases of debility. With meat nearly a shilling a pound, how can they obtain four or five shillings' worth a week ; and if it is indispensable, how are they to get well without ? But if, as is now stated—and I believe with truth—meat is principally a muscle repairer, and the force created is in reality principally obtained out of the carbonaceous food, fats and amylaceous substances, the chance of the poor is infinitely greater when " force" has to be regained. Oatmeal or any cereal with milk and oil or fat, and a small amount only of meat, will answer nutritive requirements, and a few shillings a week will go as far as ten or twenty.

I have certainly throughout my professional career remarked, as already observed, that meat-fed children and great meat-eaters are not stronger than other people. With children, indeed, I believe it is the reverse. The children whom I have attended, who have lived on meat, eating it three times a day— certainly not by my advice—have not proved as strong nor as healthy as those who have lived on a more mixed dietary. Compare these town-fed children, who eat from ten to twenty shillings' worth of animal food every week, with the Irish or Scotch peasant children, fed all but entirely on potatoes and milk, or oatmeal and milk. These researches also explain the disastrous effects which have, in many instances, resulted from the very nitrogenous or animalized dietary recently vaunted as an infallible remedy for obesity.

A problem difficult to solve in practice is the constantly observed fact of the children of really healthy parents, born and educated apparently under the most favourable conditions, dying off, one after the other, of Phthisis as they reach the adult period of life. I· have often seen large families swept off between fifteen and twenty-five, one after the other, the parents remaining alone in their old age. I think the explanation must be looked for in mismanagement in early life, in the nursery and schoolroom.

Of course, I am well aware that the advice I now

give can only be partially followed—that there ever will be persons affected with Phthisis in all classes of society by whom it must be accepted as the decree of Providence, and who must struggle with it *in situ* where it attacks them. But even in such cases, in the earlier stages of the disease, a curative treatment may be attempted by all, even by those whose means are small, or who depend on their daily labour for their bread. In more advanced disease, likewise, a lull may be taken advantage of to make the attempt. Pulmonary Consumption does not usually progress steadily, uninterruptedly; its very nature is, on the contrary, to advance *per saltum*, by jerks as it were. When not treated, it generally remains stationary for a time, then progresses, then again remains stationary, to again progress. We may take advantage of these lulls, which represent Nature's own unaided efforts to limit and control the morbid action, in order to further hygienic treatment.

We may even go further, and admit that pulmonary Phthisis has a natural tendency to limit itself, that is, to get well, if the patient does not die, by the unassisted efforts of nature. This tendency, which no doubt also exists in other chronic diseases, rationally explains many of the cases of cured Phthisis which I found in 1840 among the aged women who died under my care at the Salpêtrière Hospital in Paris, as also many of the similar instances found by all recent observers whose attention has been directed to the subject.

This tendency of pulmonary Consumption spontaneously to stay its progress and to die out, as it were, without treatment of any kind, was forcibly adverted to and described by Dr. Austin Flint in an interesting essay on "The Management of Tuberculosis,"* published in 1863.   In this essay Dr. Austin Flint gives a synopsis of sixty-two cases of arrested tuberculosis, and says (p. 359) : "The arrest of pulmonary tuberculosis, attributable in a certain proportion of cases solely to an intrinsic tendency, is a capital fact in the natural history of the disease.   It is a fact which has hitherto hardly been recognised, and certainly not appreciated.   In how large a proportion of cases this intrinsic tendency would alone lead to an arrest of the disease, we have no data for determining.   It is not easy to accumulate examples, for an obvious reason ; it is only in some occasional rare instances that the disease is allowed to pursue its course without medication or some change of habits as regards diet and regimen.   It will be long ere the analyst can gather together a sufficient number of cases to serve as a basis for establishing the natural history of pulmonary tuberculosis.   We must content ourselves with knowledge of the fact that the disease sometimes ends from its self-limitation, and that recovery takes place without any intrinsic

* "The Management of Pulmonary Tuberculosis, with special reference to the Employment of Alcoholic Stimulants." By Austin Flint, M.D. Read before the New York Academy of Medicine, June 3rd, 1863. Transactions, vol. ii. part xi.

influences. Undoubtedly it is to this intrinsic tendency
that the arrest is due, measurably or entirely, in more
or less of the cases in which hygienic measures or
medication were employed."

I would draw attention to the fact that Nature
herself, as already shown, often makes the experi-
ment for us. When an aged person dies from heart,
cerebral, or other malady, without known antecedents
of thoracic disease, and on a post-mortem examination are found in the apices of the lung, as I have so
often found them, cretaceous deposits, dry cavities
lined with thick pseudo-membrane, puckering of the
lung tissue, with firm pleuretic adhesions to the thoracic walls, there can be but one conclusion. The
individual has, necessarily, had pulmonary Phthisis
at some antecedent period of life, and has got well
without treatment, under the influence of "spontaneous limitation," perhaps combined with improved
hygienic conditions. Indeed, it is now an acknowledged pathological fact that the proportion of those
who die in old age presenting in the upper regions of
the lungs traces of past morbid tubercular or inflammatory deposits is considerable. Professor Bennett,
in his article on Phthisis, already quoted, p. 542,
says, on the authority of Roger and Bondet, whose
observations were made like mine at the Salpétrière,
" that a considerable proportion of those who die of
old age present such evidence of antecedent disease
in their lungs."

If such is the case a certain proportion of the long

lived members of the community, of those who arrive
at the natural term of life, must have passed, without
knowing it, through the early periods of pulmonary
Consumption. Instead, however, of progressing into
the later stages, in their case the disease has been
self-limited. They must have emerged from it, and
have recovered their health, without having been
aware of the danger they have incurred. Such facts
constitute a powerful argument in favour of the
curability of Phthisis in any stage of its existence,
short of the entire destruction of the lungs. If the
disease is constantly occurring, probably in periods of
temporary debility, and then getting spontaneously
well without treatment, how much more likely is
such a desirable result to be obtained under judicious
medical treatment.

Our duty as physicians is to take advantage of
these lulls, of Nature's efforts to limit and arrest the
disease, by every means in our power. The measures
we take should not be temporary but permanent.
It is with this view that I advise the young clerk,
if able, as soon as a lull takes place, or is obtained
by treatment, to give up sedentary pursuits, and
turn farmer at home, in Australia, New Zealand, or
the Cape; that I advise the young artisan to abandon
the town and to follow his calling in the country;
that I advise the town maid-servant or sempstress to
leave the city, and to find service or work in some
country place. Nearly all have country friends who
will help them in their efforts.

There was a time when, like my neighbours, in such cases, among the poor, I prescribed tonics, cod-liver oil, and a generous dietary, and thought my duty performed. Now I have learned better. I have learned to place but little confidence in the curative value of mere medicinal treatment, pursued for a time, then abandoned. If the patients, whatever their class of life, remain exposed to the influences under which the disease is generated, their fate is generally sealed, whatever the treatment. Now, therefore, I try in such cases to encourage them to make the family and social sacrifices which a more radical treatment of their disease entails. Family and social ties are as strong with the poor as with the rich, and the tendency is even stronger with them than with the better educated, to demand from the physician a mere remedy, an antidote, which is to cure their complaint without any change or sacrifice on their part. But, as I have repeatedly said in the course of this essay, no such remedy exists for pulmonary Consumption, nor is it probable that it ever will be discovered.

The various cures for pulmonary Consumption that are constantly brought forward are founded on entire ignorance of the laws of general pathology. Those who are acquainted with these laws know well how utterly impossible it is for any one of the remedies proposed, for the inhalation of any medicinal substance, or of any amount of compressed air, or for any degree of forced inspiration, to cure a disease such as I have described, one of defective and of lowered innervation and vitality.

Nothing but an appeal to the laws that regulate the development and preservation of life can have that result. An intelligent application of those laws, as demonstrated by physiology, with the assistance of climate and rational therapeutics, may, however, be made, most unquestionably, the means of saving very many lives.

During nearly twenty years I have treated my consumptive patients on the principles developed in the preceding pages. The result has been so much more favourable than that obtained in the earlier period of my professional career by the treatment then universally followed, that I am now surrounded, as I have already stated, both at home and abroad, by many patients and friends whose lives have been thereby saved. From these, the survivors of a larger body, I shall select a few cases as illustrations of the types of disease, and of the facts enunciated, and, also, as an encouragement to my medical readers. The examples thus given, as briefly as possible, will show what can be done by energy and perseverance, once we are in the right track, in the treatment of a disease even as serious and as fatal as pulmonary Consumption.

CHAPTER VII.

CASES ILLUSTRATIVE OF THE CURE AND ARREST
OF PULMONARY PHTHISIS.

In the first edition of this essay I did not give any cases to illustrate the results which I had seen obtained by the sthenic treatment of pulmonary Consumption, leaving my readers to look for such illustrations in their own practice. I have been told that a selection from my successful cases would add to the value of the essay, and I therefore append a few, beginning with my own, certainly a most instructive one. I have principally chosen cases with which I have been mixed up socially as well as professionally, and in which I have been able, consequently, to follow the varying phases of the disease, of its treatment, and of its arrest, or of its cure. As case reading is proverbially tedious, I have avoided, as much as possible, the circumstantial enumeration of local symptoms, which are much the same in all cases, as also the details of treatment. For the latter I would refer to the preceding pages, as all were treated on the principles therein detailed. Each case is intended to illustrate a pathological type or a social phase of pulmonary Consumption.

CASES OF CURE: I.—My father, a man of vigorous constitution, was fifty when I was born. He had lived well, drinking port wine daily, according to the habits of the early part of this century. He consequently had gout early in life, remained a martyr to it, and died of acute pneumonia at fifty-seven. There was no Phthisis in his family, but a decided tendency to bronchitis and pulmonary emphysema, with secondary asthma. My mother also had a vigorous country constitution, was thirty at my birth, and died at eighty-two, of cerebral congestion. Thence, no doubt, the strength element. My own constitution was, of course, unmistakably a gouty one, as demonstrated by the uric acid diathesis early developed, and by dyspepsia of the gouty type.

Otherwise healthy, I got through an incredible amount of work, mental and bodily, until thirty-eight, when I had a classical attack of acute gout, followed by obstinate bronchitis and laryngitis. These ailments, limited at first to the winter, then continued throughout the summer, existing simultaneously with symptoms of chronic gout. The constitutional health had also been lowered by repeated attacks of blood-poisoning from the imbibition of morbid fluids through the skin. Twice this occurred after post-mortem investigations, four times after the examination of living females presenting acrid inflammatory discharges. I had two such attacks whilst in the Paris Hospitals, four whilst in practice in London. The last but one occurred in 1855, and proved all but fatal. On the last occasion, after a post-mortem examination in May, 1858, I destroyed the poisoned tissues with potassa fusa as soon as I perceived the infection. I thus probably diminished the virulence of the attack, for the inflammation of the lymphatics was confined to the hand and forearm, whereas on the previous occasions it had reached the axilla, or even the walls of the chest. Shortly after this last poisoning I had a severe attack of subacute general bronchitis, which continued, with abundant purulent secretion, throughout the summer months.

Thinking my state suspicious, I consulted my friend Dr. Richard Quain, than whom no one is more competent to give an opinion in an obscure case. He examined me most carefully in August, and found no evidence whatever of lung disease, with the exception of general bronchitis. His words to me were :—"At forty-two you have an attack of gouty bronchitis such as we usually see at sixty-two. The gout is the diathesis to treat in your case." I acted on this view, which was my own, went to the Tyrol, mountaineering, for six weeks, and on my return took less wine, less animal food, and walked four miles one day, rode ten or fifteen the next. I was exceedingly busy in practice at that epoch, and could only manage the walking by leaving the carriage four miles from home, the riding by having a horse brought to me at the extreme point of my visits, and taking the ride before coming home. I now think I made a mistake in diminishing the fuel (food) and increasing the expenditure, in adding physical to mental wear and tear. It would have been all right in the Highlands, but not with the work of a large London practice.

The treatment, however, seemed to answer admirably. I gradually lost all the symptoms of chronic gout, the uric acid, the swelling and pain in the small articulations, the dyspepsia, but I did not lose the cough and expectoration. This, however, caused no alarm, for I thought it gouty, and although friends said that I looked thin and ill, I laughed at them—I felt myself so much better, so supple and free from old aches. A few months later, in February, 1859, I was seized in the night with hæmoptysis, losing at least two pints of blood. The next day, on examination, Dr. Quain found extensive phthisical deposits in the upper regions of the right lung, extending posteriorly to the angle of the scapula, anteriorly from the clavicle to the third rib. This deposit was softening in various regions. I had clearly fallen from Charybdis into Scylla. Whilst I was treating the gouty element the phthisical one had developed itself. By Dr. Quain's advice, I changed the tactics. I diminished the exercise, adopted a more generous and more animalized dietary, took cod-liver oil, threw up midwifery, and slept all but constantly at my country house, twenty miles from London. But all to no purpose; the disease continued to progress. Dr. C. B. Williams, whom I saw, confirmed Dr. Quain's statements. On my asking him whether there was evidence of disease in the left lung, he replied that no doubt there was, but that the mischief was so extensive and so advanced in the right lung that there was no longer any basis for comparison. Thus it had become difficult to speak with precision.

I thought myself lost, and had made up my mind to die in harness, but when the summer heats came the morbid deposits softened still more, hectic fever and night sweats set in, and I became quite incapable of attending to professional duties. Thereon Dr. Quain drove me away to the north, to escape the heat and for rest, and I went direct to Edinburgh, to my namesake and friend Professor Bennett. He confirmed previous opinions, told me that I had two cavities in the upper lobe of the right lung, one anteriorly large, the other posteriorly smaller; and that if the disease continued to progress, as it was then progressing, I should be dead and buried in less than a twelvemonth. At the same time he advised me to struggle for life, to give up practice entirely and immediately, and to go to the south of Europe for the winter. He added that he had seen some persons as bad as I was recover, and that it was consequently worth while to give battle to the disease. Should I be one of the fortunate ones, I might return to consulting practice in a cautious way.

This advice I followed to the very letter, and thereby no doubt saved my life. I went from Edinburgh to Loch Awe to boat and fish, whilst my friends in London were winding up my affairs. That done I departed for the Riviera, which I had previously visited as a health tourist, free from all professional ties and duties, present and future, and there settled at Mentone. For two years I entirely avoided professional occupation of any kind, spending the winters on the Riviera, the summers fishing on the Scotch lochs. I had not been three months at Mentone before I began to improve both in general health and locally. At the end of two years I was quite convalescent, and cautiously resumed consulting practice, at Mentone in winter, in London in summer.

With improvement, however, came a sensation of lassitude and prostration which showed me that I was still under the influence of the tubercular or phthisical cachexia, and this conviction has contributed to keep me out of the arena of active professional or social life. Within the last few years, to my utter astonishment, it has left me, showing what perseverance in a hygienic life can do, even after many years of disease and debility. During the early part of the nineteen years that have elapsed since I abandoned the duties of active life (1859-1878) I had ups and downs—several relapses, and a couple of hæmorrhagic attacks, like all similarly affected, however favourable the results of treatment, which it would be too tedious to dwell upon. Suffice it to say that I was for years a confirmed invalid, and that the second winter, after a dysenteric attack, caught at Naples, disease appeared in the left lung ; that as I got better, as cicatrization took place in both lungs,

I became rather asthmatic, especially under the influence of colds; and that even now I use my recently gained strength with caution.

Dr. Quain tells me there is still unmistakable evidence of past mischief, especially in the right lung. He finds in its upper region a dry, puckered cavity, into which the air enters like "air whistling through a keyhole." There is all but absence of the normal vesicular murmur in the upper lobe of this lung, no doubt from a secondary emphysematous state of the pulmonary cells around the cicatrized regions. Otherwise I am told that I have lost the look of the invalid, and I feel in fair average health for a man of my years (62), although profoundly conscious of the past lung mischief on the least exertion. A common cold brings on severe pain in those regions. I think, however, I may look upon my case, without too much complacency, as one of cured Phthisis. Since I have lived this cautious hygienic life I have seen many of my strong healthy professional friends, who pitied my downfall, succumb, because they have not had the courage I had, or perhaps because they had not a true, energetic friend, like Professor Bennett, to put the naked truth before their eyes and to encourage them to make the necessary sacrifices, regardless of their feelings (1878).

CASE II.—One of the most instructive and satisfactory instances that I have met with of arrested and cured pulmonary Phthisis is that of my friend, Dr. Crossby of Nice, which I here give with his permission. Of sound, good constitution, without antecedents of chest disease in his family, he had been carrying on a large practice, with heavy midwifery duties, for many years, in Lancashire, when his health and strength began to fail. During two winters he had repeated attacks of bronchitis and much laryngeal irritation, especially in the spring of 1859. The opinion of neighbouring medical friends who examined him at that time, and his increased ill health, caused him to withdraw from practice, and in July of the same year he went to Edinburgh to consult Professor Bennett. He was then thirty-seven years of age.

The Professor found a well-marked phthisical deposit in the upper part of the left lung, characterized by dulness on percussion, dry and moist râles. He approved of the withdrawal from practice, and advised for the coming winter a warmer winter climate than our own, mentioning the North Mediterranean coast.

Dr. Crossby did not follow this advice until November, when he had become worse. In that month, on his way to the south, he consulted Dr. Walshe in London. The latter confirmed the previous opinions as to the presence of morbid deposits in the upper part of the left lung, and strongly urged him to lose no time in going to the Riviera.

I saw Dr. Crossby about a year after this, in the winter of 1860-1 at Mentone, and examined him for the first time. His general health, he said, had very much improved during the previous year, and he had gained in weight and strength ; but he still had morning cough and expectoration, and had become liable to severe asthmatic attacks. On examining the chest I found that there was still dulness underneath the left clavicle, and in the supra spinata fossa. In these regions there were moist rattles and bronchial respiration, whilst for some distance around the dull region the normal vesicular respiration was scarcely audible. On percussion the resonance around the dull region was increased, as compared with the same region on the right side. It was clear that during the eighteen months that had elapsed since the morbid deposits had been detected, a period of care and steady treatment, the progress of disease had not only been arrested, but that nature had limited and partly repaired the mischief. The processes of absorption, cretification, and cicatrization were evidently in full operation. Simultaneously, however, with this retrograde limitation of and repair of the local disease, a remarkable change had taken place in the lung structure surrounding the morbid deposits.

The air-cells had been broken up, and emphysema of the lung tissues had supervened. This was the cause of the asthmatic symptoms, which had assumed the spasmodic form, although there had been no symptom of asthma previously, owing no doubt to his being naturally of a nervous temperament. For some years Dr. Crossby remained subject to severe attacks of asthma whenever placed under unfavourable influences in low, damp, confined situations. This tendency has, however, gradually subsided, as the physical condition of the lung has improved, which it has continuously done.

Dr. Crossby spent the winter of 1859–60 at Hyères, that of 1860–61 at Nice and Mentone, settling definitively in practice at Nice in the latter year. He has resided and practised at Nice in winter ever since. His summers have been spent in England or in Switzerland, and facing and accepting the facts of his case, he has always taken great care of himself. Under this discipline the general health gradually recovered, and has now long been thoroughly re-established. The chest disease which, twelve years ago, threatened his life, and would have probably destroyed it had he been less courageous and energetic, is indeed a mere reminiscence of the past. On the 20th of September of the year 1871 I carefully examined my friend's chest, and found, however, evident traces of the past disease. The entire left subclavicular region was preternaturally resonant on percussion, and the normal vesicular respiratory murmur was all but absent. There was evidently in this region permanent structural emphysema of the lung tissue. There was also pain over all this part of the chest whenever he suffered from catarrh or cold. This case may also be considered one of cured Phthisis. The Doctor has been in fair average health for years, and has now lost (1878) all appearance of invalidism.

REMARKS.—Independently of its illustrating the

complete and thorough recovery of health and of integrity of lung tissue in advanced Phthisis, attainable by steady perseverance, this case admirably illustrates the asthmatic complications of pulmonary Phthisis. The break-up of vesicular lung structure, in the proximity of absorbed deposits or of cicatrices, not unfrequently produces emphysema and asthmatic symptoms in cases of arrested or cured Phthisis. Many years ago, Dr. Ramadge, a sagacious and observant physician, published a work called "Consumption Curable," founded on this very fact ; only, he entirely misinterpreted its nature. Having observed clinically and in the dead room, emphysema around arrested or cicatrized lung deposits, he jumped to the conclusion that these cases had got well because they had become emphysematous and asthmatic. In reality they had become emphysematous and asthmatic, because in the retrograde process of cure there had been destruction of lung tissue, cicatrization, and puckering. This break-up of vesicular structure leading to emphysema and asthma is probably owing to a double cause— 1st, to the co-existence around the morbid lung deposits of inflammation of the bronchial tubes and of the air-cells themselves, which breaks up the latter, thus forming powerless, non-contractile air-cavities, as often occurs in acute general bronchitis ; 2nd, to the fact that the field of vesicular respiration being much diminished by the morbid deposits, cretaceous formations, and cicatrices, whilst the area of the bronchial vessels remains pretty much the same,

more air enters the remaining healthy air-cells in
inspiration than they are meant to receive; so it
acts mechanically, over-distends and breaks them
up. Dr. Ramadge, following up his pathological
mistake, advocated a preposterous method of cure for
Consumption—viz., the forcible inspiration of air into
the lungs for stated periods each day, in order artificially to break up the air-cells and to produce asthma.
He deservedly fell into disrepute, yet his clinical remark was correct.

CASE III.—In November, 1862, I was consulted
at Mentone by Mr. A., of New York, an eminent
barrister, aged thirty-four. He was sent to Mentone
and to me, by the late Dr. Elliott, one of the most
enlightened and esteemed physicians of New York,
whose premature death, from over-work and neglect
of himself, is deplored in his native country as well as
by his British friends. The patient, a gentleman
well known in literary and professional circles, had
achieved great success in life through talent and
determined energy, was one of the leading professional men of the day, and had been living many
years immersed in all the feverish cares and labours
of a busy city career. His constitution and physical
powers, however, were not equal to the duties that
success entailed upon him. The general health gave
way, he lost power, became thin, and subject to
constant cough, with morning expectoration. On
consulting Dr. Elliott, in the autumn of 1862, the
latter discovered a morbid phthisical deposit in the

upper region of the right lung, and advised him to winter in Europe.

On examination I found dulness below the right clavicle anteriorly, as low as the third rib, and also posteriorly, in the supra spinata fossa. In these regions there were moist rattles, mixed with dry crepitation. The upper region of the left lung was in a suspicious state, although there was no decided evidence of disease, beyond increased vocal resonance and prolonged expiration. He was pale and weak, with but little appetite for food. Constitution and health, originally good, although he had never been very strong. He had evidently worked more through nervous energy than on real physical power, derived from food and original vitality.

I never had a more energetic or a more intelligent patient than this gentleman. He threw all his mental powers into his case, mastered its details, and once he understood the basis on which the treatment was to be carried out, he never swerved, never made the slightest mistake. He paid me the compliment to state that he presumed I had done and was doing the best I could for myself, and that consequently he should imitate me. This he did in every respect. He took a master for Botany, acquired the groundwork of the science in a few weeks, and spent his time from breakfast to dinner out of doors botanizing, sketching, reading, in the sunshine among the rocks, without fatiguing himself. In spring he gradually ascended north—guided

by the thermometer—spent the month of June in
Surrey, those of July and August on the Scotch
Lochs. In September he descended south gradually—again guided by the thermometer—so as to
reach Mentone by the end of October. All this
time he followed rigidly the rules I have laid down
—sponging daily with cold water, living out of doors
in the day, sleeping in a well-ventilated room at
night, taking cod-liver oil and such tonic medicines
as his case seemed to require—mineral acids, iron,
quinine, preparations of phosphorus.

The very first winter there was marked improvement in the general health, and the progress of the
disease was arrested. During the subsequent eighteen
months that he was under my care the retrogression
of the disease and the consolidation of the general
health progressed uninterruptedly. At the end of
the two years the general nutrition had evidently
become quite normal. The colour was good, the
weight had increased, the appetite was fair ; sleep
refreshing, strength much improved. There was
still slight dulness on percussion, bronchial respiration, and some vocal resonance ; but there were no
rattles—dry or moist—there was no cough and no
expectoration. I thought he could safely return to
his native country, but not to active professional
life. He therefore, in conformity with my advice,
abandoned the latter, and retired to the country.
There he built himself a house in a healthy, picturesque locality, and turned country gentleman. I

have often heard from him since then, and his reports of himself have always been most satisfactory. He has continued to be free from chest symptoms, and to enjoy better health than he had had for many years before he was ill, bearing well the American winter. Dr. Elliott wrote to me some time before his death that nearly all traces of past disease had disappeared in our patient, so much so that it had become difficult to fix on the seat of the past mischief. For the last ten years my American patient and friend, after spending seven years in rural felicity, has entered into political life, undertaking important political duties. I cannot blame one so able and energetic for thus acting after nine years' health retirement, and do not prophesy evil. I had rather he had kept quiet, cultivating his acres, planting, shooting, fishing, and thus preparing for a green old age in the country. But I am happy to say that until now, 1878, he has borne his active duties well.

CASE IV.—In October, 1861, I was consulted at Mentone by Miss B., addressed to me by Dr. Little of London. He had detected the existence of pulmonary Phthisis a short time before, and wished his young patient to spend the winter in the south, looking upon the case as a very serious one; and such I found it. Miss B., aged twenty-one, had hitherto resided with her parents in a healthy country locality, in a midland county. Although delicate she had enjoyed tolerably good health until the previous spring, when it began to flag; she became thin and subject to a dry hacking cough. The summer weather not modifying these symptoms, her friends consulted Dr. Little. There were some antecedents of Phthisis in the family.

When I saw Miss B. she was pale, emaciated, weak. The tongue was white, the appetite bad, nights restless, with a tendency to night sweats. On examination I found a decided phthisical deposit in the upper lobe of the right lung, a less marked one at the apex of the left. In the right lung there was, both anteriorly and posteriorly, dry crepitation, and marked dulness on percussion, with sparse moist rattles, bronchial respiration, and vocal resonance over a considerable surface below the clavicle and above the spine of the scapula. On the left side there was only slight dulness, with bronchial respiration and vocal resonance.

This case gave me great anxiety. There were many unfavourable features : the antecedents, the early age, the marked state of general debility, the dyspeptic symptoms, the manifestation of the disease in the midst of home comforts in the country, the evident existence of morbid deposit in both lungs. These were all most assuredly unfavourable conditions, but there was one most favourable one. My young patient was meek, gentle, amiable, intellectual, confiding, and her parents were able to do all that was required. Although quite ready to die, she wished to live, more for the sake of others than for life itself. So she gave me her confidence, and quietly, but

O

perseveringly, acted up to all my requirements, to
their spirit as well as to the words themselves.
During three winters that she spent at Mentone
under my care I never once knew her do what I did
not wish her, or omit anything I wished her to do.
This can be said of but few patients.

During the three years thus devoted to treatment
—the winters at Mentone, under the most favourable
conditions, the summer at her country home in England—the course of events was chequered. The first
winter the result of all our efforts appeared uncertain.
The dyspeptic symptoms proved tenacious, difficult
to remove, and nutrition continued to flag.   All that
we gained was that the disease did not extend, that
it remained stationary. The second winter she had had
an autumn influenza cold in England, and although
I found improvement, retrogression, on the right side,
there had been, under its influence apparently, softening of crude deposit on the left. The latter was indeed
then worse than the right.   By the end of the second
winter, however, we had entered into a more prosperous and satisfactory phase.   The tongue was clean,
the appetite and digestion better, the general nutrition improving visibly.   Simultaneously the local
mischief at the apex of both lungs began to improve
rapidly.   The cough and morning expectoration all
but disappeared, the dulness diminished, the dry crepitation ceased, and the moist rattles were less general.

By the end of the third winter the general health
was quite restored, the complexion good, strength

returning, and the local evidences of disease all but gone. The following winter family circumstances prevented this young lady returning to the south. The winter, however, passed so successfully at home in England that she ceased to consider herself an invalid, and returned, although with care, to the ordinary routine of life. In 1868–9 Miss B. again passed two winters at Mentone, no longer for her own health, but as the companion of a sick relative. On examination I found both lungs quite sound, with scarcely any perceptible trace of the past. I casually met her in August, 1871, and again examined the chest, this time merely for my own satisfaction, with the same result. Many years have thus passed since I first saw this young lady, then unquestionably in a most critical state. Yet her recovery, which required three years of assiduous treatment, has been sufficiently complete to allow her return to the arena of English life, winter and summer, with all its joys and sorrows and cares, without reproducing the disease. She still remains well (1878).

CASE V.—Madame C., a young Russian married lady, aged twenty-four, was placed under my care in 1864 by Professor Bennett of Edinburgh—then himself at Mentone on a health visit—to be treated for uterine disease, which he had casually discovered. Her St. Petersburg physicians had recognised the preceding summer the existence of a phthisical deposit at the apex of the right lung coinciding with great debility and constitutional disturbance, as also

with an obstinate cough. They had sent her, therefore, to Mentone for the winter as a consumptive
patient whose life was in great danger. She arrived
early in September, and thinking that climate was all
and everything, merely pursued the treatment laid
down for her at home, as is very often the case with
our winter invalids. Professor Bennett, however, was
known to her medical friends at St. Petersburg, and
on his presence being mentioned they insisted on her
taking his opinion. He found the case such as they
had represented it—one of pulmonary Phthisis in
its early stage, a limited deposit at the apex of the
right lung, beginning to soften here and there, with
considerable constitutional disturbance.

He was struck, however, by the nausea and
inability to take food or cod-liver oil. On inquiry
he found that she had been ill ever since the birth
of a little girl, her only child, three years before ;
that she had leucorrhœa, dysmenorrhœa, back-ache,
ovarian pains, and could not walk or stand without
suffering. He then sent for me, and we found, of
course, with such symptoms, severe inflammatory and
ulcerative lesions about the cervix uteri, evidently
occasioned by the sequelæ of the confinement three
years previous. The labour had been a tedious and
protracted one. Under appropriate surgical treatment the uterine lesions soon began to mend, and,
simultaneously, the nausea ceased, the appetite returned, and we were able to initiate thoroughly the
treatment of the phthisical complication.

By spring the uterine disease was radically cured, the general health was rapidly improving, and the progress of the chest affection seemed quite arrested. Madame C. spent the early part of the summer at the iron springs of Schwalbach, the hot months on the Baltic, and returned to Mentone in the autumn.

This time she placed herself under my care from the first. I found the uterine disease cured, and all the symptoms it had occasioned were passing away. The general health had continued to improve, and the period of retrogression had commenced in the lungs. The winter was prosperous, and when she left me in the spring, to return to St. Petersburg, the cough and expectoration had all but entirely disappeared; there was only slight dulness left, no dry crepitation, and merely a few bullæ of moist rattles on forced inspiration. She intended to return for a third winter to Mentone, but family circumstances prevented it. However, the necessity had evidently ceased to exist, for she got quite well, and has had no return whatever of chest symptoms, notwithstanding the rigour of the winter in the North of Russia, where she resides habitually. I saw her in 1870, in good health, with nothing to complain of. She was at Nice with an aged relative, there for his health, and came over to Mentone to show me how well she was. Six years had elapsed since we first met. I have recently heard of her as quite well (1878).

REMARKS.—This case is a good illustration of the

complete arrest and cure of Phthisis in a young subject characterised by great constitutional disturbance, but it also illustrates the fact that such disturbance may be, in a great measure, the result of the pernicious influence of uterine complications. In Madame C. probably these complications were the first, although indirect, cause of the consumptive disease. Whilst they lasted, improvement or recovery was impossible, and it would probably never have taken place had they not been discovered and removed. The rapid improvement that took place, once the uterine disease, with all its pernicious reactions had been removed, is remarkable. This was just one of those cases that I not unfrequently meet with at Mentone, and for discovering and curing which I occasionally get into disgrace with some of my London colleagues, practising only as pure physicians.

CASE VI.—Captain M., aged thirty-two, consulted me in November, 1862, at Mentone. His medical attendants in England had pronounced him consumptive the previous winter, after an attack of hæmoptysis, and had advised him to winter in the south of Europe. He had served in the Crimea and in India, and subsequently in Canada, where he was first attacked with chest symptoms of so serious a character as to necessitate his withdrawing from active duty on sick leave. There were antecedents of Phthisis in the family. I found a very decided phthisical deposit at the summit of the right lung.

The dulness extended two inches below the clavicle, and extended over the supra and infra spinata fossæ. There were sparse moist rattles and dry crepitation over this entire region, with bronchial blowing respiration, cough troublesome, expectoration free, muco-purulent, pulse quick and small, tongue white, appetite bad, strength below par, night perspirations. He was at once put under sthenic treatment, mineral acids, cod-liver oil, cold sponging, good diet, wine in moderate quantities, life in the open air. This gentleman proved a good and intelligent patient, and acted in every respect as he was requested, although often with a subdued growl.

As is so often the case, the first winter we made little or no way, beyond arresting the progress of the disease. The summer was spent in England. The second winter there was improvement, but still the evidence of active disease remained as above, although in a minor degree. I persuaded him to sell out of the army, so as to be able to devote, henceforth, his entire time and thoughts to the struggle for life. It took above three years of constant attention, care, and treatment to establish quiescence, but by the end of the fourth year all cough and expectoration had ceased, and there were merely traces of the past mischief left. Become a man of leisure, he subsequently passed his winters in healthy localities in the south of Europe, his summers in the Alps, Pyrenees, or in England. In the spring of 1871 I met him, and examined the chest. There was no longer

any perceptible dulness over the formerly diseased
regions, no rattles, merely diminished vesicular murmur, as compared with the other side. The general
health seemed very good. He was in good flesh,
rosy, and with the hue and appearance of health.
His strength had returned, and he could ride and
gallop like other people ; indeed, he wanted to ride
with the hounds, from which I dissuaded him, merely
because he had been a dashing, daring rider in youth,
and I thought that, once on horseback again, he
might lose discretion. This I consider one of the
best cases I have seen; but again, the perfect recovery
made was due to firmness and resolution persistently
shown during nine years. I told him he could now
do what he liked in reason ; that occupation in a
healthy climate would be beneficial. I have recently
heard of him as doing well in New Zealand (1878).

CASE VII.—CASES OF ARRESTED PHTHISIS.—Mr.
D., aged twenty-two, came to Mentone October, 1863,
sent by Dr. Bowditch of Boston. He had been serving
for the two previous years in the northern armies,
during the civil war, and had encountered hardships
and privations of every kind—exposure, bad food, and
excessive fatigue. Latterly he had become greatly
debilitated, and had presented serious chest symptoms
which incapacitated him from duty. On consulting
Dr. Bowditch, the latter recognised the existence of
phthisical deposits at the apex of the right lung, and
sent him to winter at Mentone.

On examination I found dulness anteriorly underneath the right clavicle, and for three inches below, posteriorly, over the supra and infra spinata fossæ, with dry crepitation and sparse moist rattles, bronchial breathing and blowing; left apex suspicious, prolonged expiration, breathing harsh. The constitutional condition was very unfavourable. He was anæmic to an extreme degree, positively blanched, with white tongue, loss of appetite, and night sweats ; cough very troublesome, expectoration abundant. Otherwise he was a fine tall athletic young fellow, intelligent and tractable, with a sensible mother and sisters to look after him, and to keep him within bounds. This domestic companionship and superintendence is an important feature in such cases ; it is very difficult for young men left to themselves to adhere to the rigid discipline required to secure recovery.

Under the combined influence of a hygienic domestic life, of the mild winter climate, and of medicinal treatment, Mr. D. gradually recovered. At first, for a month or two, the disease seemed rather to increase than to diminish, more softening taking place. The original morbid impetus was not exhausted or arrested. Then the general health and nutrition began to improve, the tongue cleaned, the appetite returned, the blood became redder, and the complexion less blanched. Simultaneously the progress of the chest disease was evidently arrested, and towards spring decided improvement set in. The cough and expectoration diminished, and the area of dulness began to decrease. The summer was passed in the higher regions of Switzerland with decided benefit, and he returned to Mentone in October, 1864, much improved, constitutionally and locally.

That winter the favourable progress was steady, uninterrupted. When he left in May, the cough was all but gone, the expectoration reduced to three or four sputa in the morning, the dulness a mere vestige ; no dry crepitation, and only a few sparse bullæ of mucous rattle on forced respiration. The aspect had become that of health, and I considered him in full progress towards a perfect recovery. So he determined to return home, and we began to talk of the future.

Although referring him for directions to his friend and physician, Dr. Bowditch, I gave it as my opinion that he had better buy land, and become a country gentleman and farmer, which would enable him to live all day out of doors on horseback or otherwise. This advice was given, although I knew that it would, if followed, entail a great sacrifice. His father was a merchant of eminence, and as the eldest son, he of course was his natural successor. I saw him again in London, in June, looking well and in first-rate spirits, on his way to Liverpool to embark.

For two years I entirely lost sight of my young patient, when in October, 1866, he again appeared in my consulting room at Mentone. To my great regret I found him as ill as when I first saw him in 1862, four years previously. He had not followed my advice, but confiding on my report of his improvement, subsequently confirmed by Dr. Bowditch, he had resumed his position in his father's counting-house. Two years' sedentary work, two hot American summers and cold winters, had reproduced all the symptoms of the disease. There was the same blanched anæmic appearance, the same debility and want of vital power, and the same local symptoms. There had clearly been a gradual falling off in general health, and fresh deposits in the upper lobe of the right lung had formed. This time also there was distinct evidence of disease, in a minor degree, at the apex of the left, slight dulness, and sparse moist rattles, both anteriorly and posteriorly.

I was even more uneasy this time than previously. It seemed very doubtful to me whether the organism could again make a rally, again arrest and cure the phthisical disease. Very often these relapses are fatal, for there appears to be no recuperative power left. However, all the old means of treatment were again resorted to, and after a few months of uncertainty he again entered a phase of improvement and of progress—first constitutional, then local. The ensuing summer was spent in England, the winter at Mentone, and in June, 1870, Mr. D. left to return home, nearly as well as before. The war prevented his coming to Europe the next winter, but he lived constantly in the country, doing well until the year 1874, when he died suddenly of embolism of the heart following inflammation of the saphena vein.

REMARKS.—This case is interesting as illustrative

both of repeated recovery from a most unfavourable state, general and local, in early youth, and of relapse from over-confidence in the first recovery. Had this patient carried out a hygienic life in the country as did his countryman in Case III., I think it probable there would have been no relapse. He died eleven years after his first visit to me of an accidental disease.

CASE VIII.—In January, 1861, I was consulted at Mentone by the Rev. Mr. E., aged fifty-five, a clergyman from Ireland. He had been very ill, he stated, for a year, with severe cough and abundant expectoration, and had been pronounced in advanced Consumption the previous summer by Dr. Stokes of Dublin, who had sent him to Mentone for the winter. Of an energetic temperament, and under the influence of a strong sense of duty, this gentleman had always devoted himself with untiring ardour to his parochial duties. Forgetting or ignoring the advance of years, he had continued to labour in a wide sphere of activity with the same enthusiasm and self-denial at fifty as he had done at thirty, in all seasons, in all weathers. The expenditure of vital power being clearly beyond his organic resources, he became exhausted, and caught a winter cold which he could not shake off. This bronchial attack was followed by extensive phthisical deposits in the upper region of the right lung.

When I saw him there was complete dulness underneath and for four inches below the right clavicle, as also posteriorly, down to the angle of the scapula.

Below the clavicle there was cavernous blowing over a large area; around and posteriorly moist rattles and dry crepitation, with bronchial blowing. Evidently the upper third of the right lung was infiltrated with phthisical deposits, and anteriorly there was a large cavity. The left lung was comparatively sound, but the apex was suspicious, respiration harsh, voice resonant. There was a good deal of chronic bronchial irritation of the larger bronchial tubes in the right lung, but little in the left, which showed that it was connected with and the result of the phthisical deposits.

This gentleman had none of the querulousness which often accompanies chronic disease in advancing life. He was calm, impassible, intellectual, accepted his malady and its symptoms, and bringing an exceptionally clear intellect to his assistance, helped me in every possible way. The result was a slow but gradual and progressive recovery. The first winter he had several attacks of hæmorrhage and the disease was stationary. The second, his health was better, he had no hæmorrhage, and improvement commenced. The third, the retrograde stages were rapid, and by the end of it full convalescence was established. All secondary inflammatory conditions had ceased around the lung deposits. Part of the latter had softened and been evacuated, part had no doubt become consolidated and cretaceous, and the cavity had much contracted. There it was, however, secreting mucopus in small quantities, and still giving rise to moist râles and cavernous respiration. The summers had been spent at home in Ireland. By this time my friend's health appeared quite re-established to a superficial observer; he was rosy, and had the hue of health, although in reality an invalid. Since then Mr. E. has spent most of his winters in the south, at Mentone, Palermo, in Corsica, and elsewhere. One winter he tried staying at home, but had an attack of acute pneumonia at the base of the left or sound lung, which nearly killed him. However, he surmounted it, and when I saw him the subsequent autumn, there was no trace left—a fact which shows the great difference there is between real pneumonia and pulmonary Phthisis. Ten years had elapsed, in 1871, since I first saw Mr. E., then apparently on the brink of the grave, 1861. He was then sixty-five, healthy, enjoying life as a rational invalid. He always returned to Ireland in the summer. I have not seen him since.

CASE IX.—In December, 1862, I was asked to see a young French lady, Miss F., aged twenty-one, who was said to be dying of decline under homœopathic management. Although rather unwilling to interfere in a case considered hopeless under such circumstances, I consented to see her, called, and found her apparently dying of advanced Phthisis. She had taken leave of her friends, they of her, was given over by her medical attendant, and was preparing for impending death. The whole of the upper lobe of the left lung was infiltrated with phthisical deposit, there was a large cavity under the clavicle, constant cough and expecto-

ration, night sweats, diarrhœa, emaciation, inability
to take food, pulse small, a hundred; in a word, all
the symptoms of the last period of the last stage of
the disease. I really felt myself that the case was
hopeless, and was reluctant to take her out of the
hands of people she liked, merely to see her die in
mine; but then there was the reflection that under
homœopathic treatment nothing whatever was being
done medicinally even to mitigate symptoms, whilst
the hygienic management of the case was simply
atrocious. She was, as is so usual on the Continent,
lying in a small, warmed, closed room, in a really
pestilential atmosphere. The perspiration was scarcely
wiped off the body, for fear of a chill, and she was
taking slops only.

So I accepted the responsibility, and overpowering
opposition on the ground that as she was said to be
dying it could not matter what changes were made
in the treatment, I at once put her under a totally
different system. Internally I gave astringents and
mineral acids to arrest the diarrhœa and perspiration, coupled with eggs, milk, fowl, fish, and wine,
instead of the slops she was taking. Then I opened
the window, got pure air into the room, day and
night, and had her placed in a hip-bath before the
fire, and sponged over, at first once, then twice a day,
with water ·at 70° Fah. All this was done in mid-
winter, to the intense horror and dismay of the family
and friends, but in less than a week, to their surprise,
and, I must add, to mine, a complete change came

over the state of things. The diarrhœa and perspiration stopped, the tongue cleaned, and the food was both retained and digested. Gradually the hectic fever became less, the stomach was able to bear cod-liver oil, with acids and bitters, and she began steadily to recede from death. By the month of May the young lady was plump and fat, in very tolerable general health, able to walk a mile daily, sleeping and eating well, and only coughing and expectorating on waking in the morning. The diseased lung was also in a much more satisfactory state, the morbid symptoms gradually subsiding, and becoming limited to the vicinity of the cavity. She came two more winters to Mentone, spending the summers in the north of France.

After three years thus spent she had become apparently well, rosy, plump, fresh-coloured. The lung mischief was reduced to a moist cavity underneath the left clavicle, with some moist rattles around, and diminished sonority on percussion. Since then my young friend, whom I have seen most years in passing through Paris, has had to enter the arena of active life, has had many cares, many sorrows, has been obliged to sacrifice herself to home affections, to go out in all weathers, and to endure fatigue beyond her strength. On several occasions irritation has set up around the cavity, threatening to renew the active state of the disease ; and she has had several attacks of hæmorrhage, all traceable to its walls or to the region of the lung around the cavity. But with all this she holds her ground (1878), and may fairly be looked upon, after sixteen years, as a case of arrested, if not of cured, Consumption. Had she been able to take that care of herself that the first six cases did, the recovery would probably have been as complete as it has been with them.

An important feature in this lady's history is that a year before she was taken ill a brother, one year her senior, the only other child of her parents, died in a few months of rapid decline. He was treated by M. Bouilland, of Paris, on his well-known anti-phlogistic principles, by repeated depletion through leeches and a low starvation diet. The sister would have unquestionably died in the same way had I not come to the rescue with rational therapeutics and a sthenic treatment.

CASE X.—In December, 1860, I was consulted respecting Miss G., aged twenty-two, by her mother. She was born and educated in a large manufacturing town in the north of England, and had been pronounced consumptive the previous summer. Her father and mother, brothers and sisters, were healthy, and she herself had been so until about a year before, when her health and strength began to flag. Being of an active, energetic frame of mind, she had for several years over-exerted herself, both mentally by study and physically in district visiting and works of charity. On passing through London, a couple of months before I saw her, an eminent physician, whom her mother consulted, drew her aside and told her that her daughter's case was very serious, and that she must make up her mind to lose her, in all probability, before the winter was over.

On examination, I found an extensive phthisical deposit in the upper region of the right lung. The dulness descended anteriorly three inches below the clavicle and into the infra spinata fossa posteriorly. There was considerable sparse softening, troublesome cough and expectoration, appetite bad, tongue white, great lassitude, but no emaciation. On the contrary, the patient was rather plump, and had not the appearance of a consumptive. The left lung appeared sound. In this instance, again, both the mother and the daughter were sensible, rational people, and did all they possibly could to help me. At first the disease continued to progress, and a small cavity formed under the right clavicle. At the end of the first winter it became stationary and the general health improved. So not only did she not die, but returned home better. The second winter decided retrogression commenced, and by the end of the third, complete convalescence had set in. Absorption, cretification, cicatrization had taken place, until there was only a small cavity left, probably filled with mucus. This, with slight induration of lung tissue and some thoracic adhesions, were the only evidences of the past.

These three winters were passed at Mentone. Subsequently Miss G. spent a winter at Torquay, which was not successful, as she suffered much from bronchial irritation. So she again came to Mentone, and continued to do so until her death. In the year 1868 she had a very severe attack of continued fever, typhoid, which nearly cost her her life. She rallied, but subsequently a phthisical deposit formed at the apex of the left lung, which went through the same phases of softening and cicatrization as had done that on the right. Although terminating thus favourably, it left the same traces as on the right—a small mucus-filled cavity. Over an area as large as half a crown there were heard a few moist bullæ on deep inspiration, always in precisely the same region. There was no perceptible dulness on either side. This lady's natural energy enabled her to take an active part in life—perhaps too active—for some years. In 1872, however, laryngeal Phthisis set in, slowly but surely progressed, and brought her valuable life to a close in 1874. In this case life was prolonged fourteen years, and would probably have been permanently saved, had it not been for the accidental attack of typhoid fever from which she never thoroughly recovered.

CASE XI.—In November, 1861, I was consulted by Mr. H., aged twenty-one, from the south of England, who had been declared consumptive and sent to Mentone by his own medical attendant, with the sanction of several London physicians. I found extensive phthisical deposits in the upper region of the right lung, and a cavity underneath the clavicle anteriorly, with the usual symptoms, local and consti-

tutional. There was great emaciation and debility.
The family health was good. Under sthenic treatment, the disease evidently ceased to progress, was
arrested the first winter, but there was very little
outward evidence of improvement until the middle
or end of the second, which was also spent at Mentone. The intervening summer was passed in England in a cool locality. During all this time Mr. H.
presented a peculiar and unusual feature. Every
morning on waking, the mouth filled with blood on
the first efforts of coughing. The blood evidently
proceeded from the diseased lung regions. At first
this slight hæmorrhage caused me anxiety, but we
gradually got accustomed to it, looking upon it as
beneficial, as nature's mode of relieving congested
vessels. Hæmorrhage in Phthisis often really has
this character. The hæmorrhagic tendency lasted
about two years, during which time we had several
more severe attacks. They always gave way, however, to treatment.

By the end of the third year Mr. H. was quite
convalescent, the lung mischief being reduced to a
moist cavity underneath the right clavicle, and to
the traces of past lung disease already so often
described. From that time Mr. H. continued to
spend his winters in the south of Europe and his
summers in England, in a very bearable state of
invalidism, but not in perfect health. His recovery
would probably have been more complete had it not
been for several attacks of acute pleurisy which

came on after rather imprudent exertion. They proved difficult and tedious to subdue, leaving for long false membranes, and permanent adhesions. As a result of these morbid conditions a good deal of emphysema, with increased resonance on percussion, and diminished vesicular murmur showed itself in different parts of the lungs. This gentleman was one of the most cultivated men I ever knew. During many years he did all in his power to further my efforts. He continued to enjoy life and to devote his time to scientific pursuits, at Mentone in winter, at Kew in summer, until the autumn of 1875, when he died of acute bronchitis after a three weeks' illness. His life was thus prolonged fourteen years.

CASE XII.—In October, 1865, I was consulted at Mentone by Mr. L., aged 28, an author and University man. This gentleman had studied very hard during his University career, as also subsequently, consuming the midnight oil, and otherwise neglecting his body for the sake of cultivating his mind. The previous spring he was attacked with a troublesome cough, followed by hæmoptysis. It was then discovered that he had phthisical disease at the apex of the left lung. This opinion, given in the country, was confirmed in London, and Mentone was fixed upon for a winter residence. I found extensive tubercular deposits at the apex of the left lung, then softening, and crude deposits of a more limited character at the apex of the right. On the left side, the dulness descended anteriorly to the level of the heart, and there were moist and dry rattles over the entire region. Posteriorly the dulness reached the spine of the scapula. On the right side there was merely slight dulness in the supra spinata fossa, with bronchial respiration and prolonged expiration. The digestive system was very much out of order, tongue white, appetite bad, liver not acting, and digestion laborious. I was long very uneasy about this patient. The digestion remained defective all winter, and he had repeated attacks of haemorrhage. At the end of the winter all I could say was that he was no worse than when he arrived.

He returned to England in summer, and here he had a severe attack of fever, probably the result of softening of morbid deposits in the right lung. He was very ill, and the physician whom he saw in London, on his way through, thought him decidedly worse than a twelvemonth previous. When I examined him at Mentone, I did not share this opinion. I found there had been, clearly, softening on the other, or right side; but on the left side, the disease was decidedly stationary. Along with the secondary softening of the deposit at the apex of the right lung, possibly as its cause, there had been all but general bronchitis, from which he was still suffering. It was this circumstance that made me take a more favourable view of the case. Once a phthisical deposit exists, it must soften and be evacuated if it is not absorbed. I often remark that when the process of absorption and limitation is decidedly going on in one lung, under some influence or other old deposits or new soften in the other. This winter proved as trying as the former one, from cough, expectoration, repeated hæmorrhages, and dyspeptic symptoms. By the end, however, we had gained ground ; quiescence was getting established in both lungs, the cough diminishing, digestion better, and strength returning.

The second summer was again spent in England under much more favourable conditions, and a third winter was initiated at Mentone. This time my patient came at his own express desire. The hæmorrhages had been so frequent that I myself was afraid the air might be too keen, and told him he might try some moister climate. But he liked the place, thought that he was better, felt that sometimes the hæmorrhage had actually done him good, and insisted on returning. This winter all changed for the better. There was no hæmorrhage, the digestion became good, the general health and strength rallied, the cough and expectoration became trifling, and all the local symptoms rapidly diminished. When he left he was in a fair way to recovery. The summer was favourable, and the next winter was passed in England, with most satisfactory results. Indeed the perseverance and courage of this gentleman have, I believe, all but landed him among the cases of cured Phthisis. I hear that he continues well (1878). He has always passed the winter in one of our English sanitaria since the year 1869.

REMARKS.—This case illustrates, forcibly, the dys-

peptic form of Phthisis occurring in a literary man
of studious habits.   He made no real rally until the
digestion improved, when the improvement was very
rapid.   The local disease was clearly subordinate, as
usual, to the constitutional state.   It also illustrates
the difficulty of deciding whether repeated hæmorrhage from the lungs is merely a symptom of the particular form of disease under observation, or is a
result of unfavourable climatic influences.   This patient, like Case XI., suffered from hæmoptysis for
two years, but ceased to experience it as soon as he
began to get better, although in the same locality.
Had they changed their abode the third winter, the
cessation of the hæmorrhage would have been attributed to the change.   The case is a good example of
progressive recovery in England after three winters
passed abroad.

This gentleman, like the eleven other patients
whose cases are narrated, was intelligent, rational,
tractable, persevering.

Indeed it is a remarkable fact that among my successful cases of the treatment of Phthisis—of which
those given are only a selection—I do not find a
single foolish, wilful, obstinate person.   Folly is a
sad, a lamentable mental condition, one as detrimental to the retention or to the recovery of health
as it is to wordly prosperity..   In pulmonary Consumption, when the disease has entered the progressive stage, it is all but equivalent to a verdict of
death.   That such should be the case will be easily

understood by all who analyse the cases narrated above. In every single instance the disease was only arrested in its progress or cured after years of obedience to advice, after years of self-control and self-denial, extending to the wishes, instincts, impulses, passions. Those who from shallowness of intellect, or from perverseness, from folly in a word, cannot or will not act thus have scarcely a chance in the struggle for life. The sight of a succession of such people, making a feeble effort, miserably failing, and then succumbing, one after the other, makes the experienced physician understand the withering anger and contempt shown by Solomon, the wisest of men, when speaking of fools :—" Though thou shouldst bray a fool in a mortar among wheat with a pestle, yet will not his foolishness depart from him."—Proverbs.

CHAPTER VIII.

WHAT ARE CURED CONSUMPTIVES TO DO IN LIFE?— CAN THEY MARRY?

THE cases narrated in the preceding pages must naturally lead those who peruse them to ask the questions at the head of this chapter :—First, What are those in whom pulmonary Consumption has been cured or arrested to do in life? Second, Can they marry? I will endeavour to answer these questions, to the best of my judgment, as briefly as possible.

The illustrative cases given, as it will have been perceived, are divided into two series. In the first six the disease, to all outward appearance, was cured, and the patients have recovered a fair average state of health. In the remaining six its course was arrested during a long sequence of years, but complete local quiescence was not obtained, and the general health was not thoroughly re-established. Those who belong to the first category, would certainly be justified in returning to the arena of active life, and might live to die of some other disease, or of old age. I would incidentally remark that dying of old age means dying " in old age" owing to exhausted vitality, but often of a morbid state, which in a younger and stronger person would not produce death. In 1840 I made several hundred post-mortem examinations at the Salpétrière, and only in two or three did I fail to find morbid conditions which satisfactorily accounted for death. I consider that really cured Consumptives who return to active life, may be compared to some of those aged women in whom I found, after death from other disease, the traces of past pulmonary Consumption.

The younger the patient is, the more thoroughly the general health is re-established, the more likely he or she is to pass out of the phthisical diathesis. At the same time it is decidedly more prudent, more judicious, even for thoroughly cured Consumptives, not to return to active life, unless socially or morally obliged so to do, as many no doubt are. If possible, it is far more prudent, more judicious, to accept the warning, as I have done, to abandon the active, ambitious careers of life, and to be satisfied with a more humble but more healthy and more hygienic sphere of activity. This applies more especially to young men, as I have already stated, whose occupations can be easily changed. Men more advanced in life often cannot alter or change their career, and to them the question must generally be one of retiring on half or quarter pay. But even that is better than a possibly fatal relapse.

Very few indeed can calculate on being as fortunate as Portal, the celebrated French pathologist, who lived at the end of the last century and at the beginning of this. In early life, at the age of twenty-five, he was pronounced Consumptive, and spent two winters at Montpelier, then a favourite resort for Consumptives, and the summers in the north of France. He got quite well, entered into practice in Paris, acquired great fame in the treatment of insanity, made and lost three fortunes, and died at the age of seventy-five, in harness up to the last. His voice had been permanently modified during the illness of his youth, and he maintained, throughout life, that he had been Consumptive and had got well. As however percussion and auscultation were unknown in those days, the diagnosis had been merely a matter of opinion, and his friends always insisted that he had had nothing but bronchitis. On his death-bed he requested three friends—M. Serres, M. Clement, and M. Gendrin—to make a post-mortem examination in order to clear up the question. This was done (1822), and all the traces of past but cured phthisical disease enumerated above were found, in a most irrefragable manner, fifty years after the cure. These details I had from my old friend M. Gendrin, who, with M. Serres, examined Portal according to his wish.

With reference to the second class of cases—those in whom the disease is only arrested, and in whom the general health has not been re-established—there can be but one opinion. They are not fit for active life—for its cares, anxieties, duties, and labours—and should keep out of it if they can by any possibility

do so. They must strive to pass into the first category, and if they cannot accomplish it, they should
bravely accept invalidism, with all its drawbacks,
and endeavour to make their invalid existence as
pleasant to themselves and as useful to others as they
possibly can.

In answering the second question—Can Consumptives marry? we may make the same distinction between those in whom the disease appears cured, who
have recovered their health, and those in whom it is
merely arrested. In the first instance marriage may
be admissible, as is active work—although there
are dangers therewith connected; but in the second,
with those in whom disease is only arrested, marriage
should certainly be avoided entirely—put off until
sounder and happier days.

The dangers of matrimony for Consumptives differ
according to sex. In men there is often an absence
of proper discretion, and consequently marriage is
an additional cause of debility and exhaustion. In
women the danger is different. Marriage does not
with them try or fatigue the constitution to any great
extent, unless followed by uterine disease. But then
it is generally followed by pregnancy, inasmuch as five
women out of six are fertile. Thus the Consumptive
will most likely have to encounter pregnancy, confinement, and nursing, with all the attendant shocks
and drains on the vitality. As a result pulmonary
Phthisis is constantly accelerated and rendered fatal.
I have seen very many cases of the kind, and consider

pregnancy a most unfavourable complication to the disease. I always dread and deplore its presence. No conscientious physician, therefore, ought to sanction a Consumptive female marrying in whom the disease is not quite arrested and cured. In married life child-bearing should cease if the wife is Consumptive.

In the above remarks I have only spoken of the Consumptives themselves, but we cannot disregard their progeny. What of them? My experience teaches me that children born of actually consumptive parents are deficient in vitality. They are born tainted with constitutional weakness, and often die teething, from tubercular meningitis, from ordinary children's diseases; or they merely grow up to become victims of the same malady as their parents.

The latter cannot give them what they have not to give—vitality, strength, health—so their span of life is short. The Fates weave them but a short thread, which soon comes to an end. Not but that, through soundness of health on the side of one parent, and by means of hygienic nurture, education, and occupation, even in such cases life may often be prolonged to its normal term. Thus hope is still left.

The practical bearing of these facts, if true, as I believe them to be, is obvious. A young man with a tendency to Consumption, or a Consumptive who has recovered, once he has emerged from the phthisical diathesis, may marry and have children who may be strong and live. But he must marry a young, healthy woman, of a good healthy stock, born and bred in the country; he must be discreet in married life, and he must bring up his children hygienically, in the country, devoting them to country pursuits. Whilst he is still acutely consumptive, it is an act of folly and cruelty for him to marry, thereby exhausting his ebbing vitality, bringing miserable diseased children into the world, and turning his helpmate into a nurse.

In southern Europe there is a general conviction that pulmonary Consumption is infectious, and this belief is, there, an additional argument against the marriage of Consumptives. Some of our medical brethren share this conviction. As I have already stated I do not believe in the infectious character of the disease, but I have a strong belief that living in the pestilential atmosphere in which all Consumptives were formerly made to live, and in which many are still made to live, day and night, is very likely to develop the disease in the healthy. Thus, I think, is explained the cases, not very rare, of wives becoming consumptive after attending their husbands, or of husbands becoming consumptive after attending their wives. To this influence must be added, in most cases, great mental depression.

With the young female in the same state the danger is much greater, as we have seen, for one or more pregnancies may be expected, which may, and very likely will, fatally precipitate the issue of events.

In the actual practice of life, however, I find that all these considerations have little or no influence over persons' actions, unless it be in the case of the very young, under the absolute control of reasoning parents. Consumptives marry, and will continue to marry, I believe, just like other people, consulting affection and worldly considerations, and showing supreme indifference to medical vaticinations. With some of the gentlest and best of each sex, dire disease in the one they love is merely an additional inducement to marriage. They long to devote their lives to the object of their affections, and no vista of disease, suffering, and death, terrifies them.

Such being the case, the natural and divine laws which regulate the well-being of the earth and its inhabitants, regardless of their wishes and actions, come to the rescue and prevent the "degeneracy of the race." As is the parent, so is the offspring ; like begets like. The diseased parent begets diseased children, who, not being fit to continue the race in its integrity, die off like plants that perish before they blossom and seed, and the earth remains the heirloom of the strong. Viewed in this light, Consumptives may and do marry, and probably will marry, as others ; thus enjoying the affections of conjugal life and of paternity like the rest of the community, albeit for a short term.

Is a short life, however, philosophically, such a very
great calamity? The statement made by the Nestor
of Medicine, "Vita brevis, ars longa," is literally
accepted as a truth. But is the first part literally
true? Is human life "short" when prolonged to its
ordinary healthy duration, three score and ten years?
I have often thought that this part of the axiom, if
thus taken literally, is utterly false. A life thus prolonged is a very long one indeed, as compared with
the lives of the animals around us, and of most of the
vegetable productions of the earth. Even if we measure it by political events, by social changes, what a
long series of both does the memory of the man who
can look back fifty years run over! How many winters he has seen, how many harvests he has helped to
consume! Even a child that dies at eight or ten has
lived the entire life of a domestic animal; infancy,
youth, middle-life, old age. Under favourable circumstances, also, the child has had a happy, joyous
life, free from cares and anxieties. Does the father
or the mother, who through this child have known
the pleasures of paternity and maternity, regret its
existence?

Thus, even if consumptives discard our advice and
marry, unwilling "nec propter vitam vivendi perdere
causas"—all comes right in the end, and the human
race does *not* degenerate.

Does the marriage of blood relations tend to
degeneracy, to imperfect mental and bodily development, to idiocy, insanity, scrofula, phthisis, as is

constantly asserted? I believe this is only true when a morbid taint exists on both sides.

All over the world are found, in remote places, healthy populations that have intermarried for ages. I have a property in a mountain village near Mentone, Grimaldi, where the 400 inhabitants are all blood relations, have intermarried for centuries, have nearly all the same name, and nearly all die in old age.

The same fact is observed in all mountain regions, in all sparsely inhabited countries, all over the world. Were the intermarriage of blood relations to be invariably followed by degeneracy, the tribes and communities so situated would gradually become extinguished. They do not, evidently because the stock is a healthy one.

The case is very different when marriage takes place between blood relations debilitated by civilisation, by town influences, and by the constitutional and hereditary cachexiæ thereby engendered. The marriage of two such individuals intensifies the morbid taint, and no doubt leads to the results above enumerated.

APPENDIX.

I. The Winter Sanitaria of the United States of North America.

II. Switzerland and its Mountain Sanitaria.

III. The Balearic Islands, Majorca and Minorca.

I am aware that the above addenda would appear more appropriately in my climatology work, " Winter and Spring on the Shores and Islands of the Mediterranean," but I crave the indulgence of my readers on the following grounds :—

The last edition of the work alluded to, the fifth, being tolerably complete, a large number of copies was printed, so that it will be some time before it is exhausted, and the journeys described have been made since its publication. All questions of climate, however, have an interest to those phthisical sufferers who are able and willing to leave their own country in search of a better one for winter or summer, and I think the information I give will prove interesting to them.

The few pages on American Sanitaria, derived from a good authority, throw a vivid light on the health resorts of the United States.

My exploration of Swiss mountain sanitaria will help to guide those who seek in Switzerland a refuge from Mediterranean heat in summer, or who wish for mountain air, whencesoever they come.

Q 2

The section on the Balearic Islands opens out all but new ground, as these Islands are but little known, although we occupied Minorca during a great part of last century. I could find nowhere an account that satisfied me, that could guide me, in appreciating their value as health resorts in winter, and this was my principal reason for going there. Were they a " Spanish Eldorado" better than the main land? My descriptions were written on the spot, and appeared last year in a horticultural journal, which will account for their botanical and horticultural character. I have, however, retained them as they first were written. To those who are unacquainted with my large work they will, I think, convey a truthful idea of what a Mediterranean island is in spring, one of the most charming sights in nature. It shows them, at least, where they should *not* go in winter. I have given at the end the conclusions to which I came, and the data on which these conclusions are founded.

I.

The Winter Sanitaria of the United States of North America.

Dr. George Walton, of Cincinnati, Ohio, has sent me a pamphlet entitled " A Comparison of the European and American Climatic Resorts," Philadelphia, 1877, with a valuable chart, which comprises in a few pages a vast amount of information. I here append, for the benefit of my readers, British and American, the conclusions at which he arrives :—

" On a general view of the temperature chart, one is likely to be impressed with the fact that the European resorts are much more equable than the American. Whilst the mean monthly ranges of the most popular European winter stations lie between eight and eighteen degrees, those of the American resorts fall between twenty-three and forty degrees, with one exception, that of Nassau, Bahamas, which ranges between nine and seventeen degrees. This we would expect to be the case, as it is well known that the entire North American continent is subject to extreme fluctuations in temperature, which are unknown in Western Europe. The cause is not far to seek. The trend of the mountain ranges throughout this country is from north to south in parallel lines, forming broad avenues, up and down which the winds sweep in unimpeded course from the equator to the pole; whilst in Europe the mountains lie east and west in frowning battlements, which roll back the fierce north winds on the plains of Germany and France, or toss them high in the heavens, to be lost amidst the warm winds from the Great Desert, leaving the winter stations which nestle on the north shores of the Mediterranean untouched by their frigid breath. Another influence tending to equalise the temperature of the European resorts is the vast land-locked Mediterranean Sea, upon the coasts of which the European stations mostly lie. The effect of the evaporation of such a vast volume of water in absorbing latent heat—thus lessening the furnace-like blasts of the Sahara, or by condensation of its moisture by the north wind throwing forth this latent heat on the surrounding shores—need only be mentioned. The Gulf of Mexico, on our coasts, does not compare with the Mediterranean in this regard, for the current of its warm waters flows directly away from our coasts, across the Atlantic; and the western winds, which, according to Blodget, prevail there during the winter months, sweep much of the beneficial effects of its moisture-laden air far out to sea.

"When we direct our attention to the rainfall, we perceive that the European stations are much more subject to rain than those of our own country. The majority of the stations in this country average between one and two inches per month, whilst those of Europe range from two to four inches per month. However, we are not to conclude from this that the humidity of the air at European resorts is

greater than at the American ; for it is a well-known fact
that the atmosphere may be heavily laden with moisture,
and yet little precipitation occur. Neither should we conclude that a considerable precipitation indicates a great
deal of cloudy weather, as in tropical showers a large
amount of rain may fall in a few hours. For this reason a
chart or table showing the average number of rainy days
for each month is desirable ; but the facts necessary to con-
struct such a chart are not obtainable.

" From generalities we proceed to particulars. The European resort which of all others is in greatest repute is that
of Mentone, possessing a temperate, equable, and comparatively dry climate. What are the characteristics of that
climate, and which of our American stations most nearly
approaches it ? We find during the cold months, from
November to April, that the average monthly temperature
of each month falls between 47¾° F. and 58½° F., and with
this an average monthly range of between 10° F. and
15° F. ; that is to say, that at no time during the winter is
the thermometer likely to fall below 40° F., or go above
70° F. Frost may be said to be unknown ; at least it
occurred only once in twenty years, and this in a latitude
43° north—3° north of New York. We need hardly a
reminder that such a climate is invigorating to persons in
health, and also for those suffering from disease who have
sufficient vitality for reaction. Which of our American
resorts is most nearly like Mentone ? We answer Aiken,
South Carolina, whether we consider mean monthy temperature, mean monthly range, or average rainfall. But
there is a marked difference in the range. While the mean
monthly temperature of Aiken is but 4° F. below Mentone,
yet we find there a monthly range varying from 28° F. to
37½° F. during the winter, a range which insures the occurrence of frost frequently, and causes the invalid an infinite
amount of care in adapting himself to these changes. The
differences in the physical characteristics of the places need
not be entered into, as they are quite well known. Nice

is very similar to Mentone in the mean monthly range, but
is subject to sudden changes, in this respect quite closely
approaching Aiken. Pau, which is also a much frequented
winter resort, especially patronised by the English, has no
analogue among American resorts. It averages five degrees
colder than Aiken, and is quite as changeable. It is also
a rainy and misty place. It is situated amidst the mountains of the Pyrenees, on a plateau seven hundred feet
above the sea level.

"Malaga probably possesses the most favoured winter
climate of all Europe, and were the hygienic conditions of
the city and surroundings more favourable, would undoubtedly be the most popular resort. The mean monthly
temperature from November to April, inclusive, lies between
$54\frac{1}{2}°$ F. and $62\frac{1}{2}°$ F., whilst the monthly range is very small,
not over 8° F. to 12° F. the entire season. Algiers, which
looks towards Malaga from the north shores of Africa,
possesses a similar climate, though not quite so equable.
Both climates are well adapted to those who shrink from
moderate cold, who are troubled with continual chilliness,
and who feel better in summer than winter. What are the
corresponding climates in this country? These are to be
found on the coasts of Southern California—Santa Barbara,
San Diego, and adjacent places. While all these stations,
Malaga, Algiers, Santa Barbara, and San Diego, only vary
from each other some 5° F. in the mean monthly winter
temperature, yet we have the inevitable fickleness of our
American climates to mar the analogy. The mean monthly
range of the European places falls between 7° F. and 12° F.,
while that of Santa Barbara and San Diego lies between
27° F. and 41° F., Santa Barbara being much the most
equable of the two. Thus whilst the average monthly temperature of Santa Barbara and San Diego for January is
about 53° F., and Malaga and Algiers about 55° F., yet at
the American places the thermometer may, during the
month, be as low as 38° F. or as high as 72° F., whilst at
the European places it would not go lower than 47° F. or

higher than 68° F. As to clear sunshiny weather, these
American stations are in every way equal to the European,
and with care will undoubtedly yield equally favourable
results in the treatment of disease.

"Lastly, we come to the Florida climates—under-estimated
by some and over-estimated by others. Typical of these
stations are Palatka and St. Augustine. With what
European stations may they be compared? None, unless
it be Madeira and the Canaries. As to mean monthly temperatures, Palatka and St. Augustine approach Madeira
within about 6° F. But, when we turn to the other side of
the picture, and look at the mean monthly range, we find
Madeira between 11° and 13°, whilst St. Augustine is
between 23° and 31°—a range for the Florida resorts which
will permit the temperature to fall inconveniently near the
frost line. As to the rainfall, the figures are favourable to
Florida, and if a series of hygrometric observations were
made both at Madeira and the Florida resorts, the results
would doubtless favour the latter places.

"Nassau is quite isolated. The climate is tropical in
character; with a very high mean monthly temperature
during the winter months—between 73¾° F. and 79° F.,
there is a small monthly range—from 9° to 18°—in this
respect more nearly approaching the European stations than
any places on this side of the Atlantic. The rainfall is
also moderate. A perpetual summer prevails without the
inconvenience of intense heat.

"Kingston, Jamaica, is given on the chart only for comparison. The city is unfitted for invalids. There are said
to be places on the ranges of Blue Mountains in the interior
of the island where the average winter temperature is 64° F.,
exceedingly equable, dry, and bracing.

"The only places with which Denver, Santa Fé, and
St. Paul could be compared are the mountain resorts of
Switzerland. For those places temperature tables are not
obtainable.

"In conclusion I may say, that when we consider the

dry and bracing air of the plateaus of Colorado, the warm and cooling atmosphere of the resorts of Florida, the Italian-like skies of Southern California, there is little reason why we should send patients across the Atlantic, far away from home and friends. With proper care all the benefits from change of climate can be had in this country that can be obtained elsewhere."

I have reproduced Dr. Walton's statements textually, including his conclusion. I must, however, add that I do not think his premises warrant the latter. The fact of physical geography to which he draws attention, that the mountain chains of North America all run longitudinally from north to south, and allow winds from the north or the south to course through the continent without let or hindrance, points to a great difference in favour of the Mediterranean basin. Owing to this physical conformation of the American continent, and to its inevitable results, all American continental sanitaria, as Dr. Walton states, are liable to great variations of temperature. The means of maximum and minimum temperature are much wider apart than in the sanitaria on the shores of the Mediterranean. The winter climate is more equable in the latter, and thence, no doubt, the favour with which the Mediterranean is viewed by many of the leading physicians of the United States.

Dr. Walton speaks of Nassau (lat. 25° 5') with its semi-tropical climate and vegetation, where the monthly range is only from 9° to 18°, with a high monthly temperature during the winter months, between $73\frac{3}{4}°$ and 79° Fah. He does not, however, allude to the Bermuda Islands, situated several degrees more north (lat. 32°) in the same seas, 600 miles from the nearest land, Cape Hatteras, North Carolina.

These islands belong to England, have an English Governor and garrison, and present all the comforts of an old English colony—roads, churches, schools, hotels, and shops. The climate is much cooler than that of Nassau. Between December and March the thermometer is said to range from 60° to 66° only. The vegetation is sub-tropical.

From the position of these low coral-formed islands or islets in mid-ocean, 500 in number, extending over a space of only twenty miles in length by six in breadth, there is necessarily a deal of wind, with frequent rain, in winter. Yet the winter climate is described as very delightful and salubrious, even the rainy days being generally fair in part. It appears to much resemble Madeira in meteorological and health characteristics, but has the advantage of being English. There is a weekly mail-packet from Halifax, Nova Scotia, and a fortnightly one from New York, which latter reaches Bermuda in four days. The islands being all but level, do not present the glorious mountain scenery of Madeira; but to invalids this is an advantage in a mid-ocean island where the protection of mountains is not wanted. The roads being good, and the walks, rides, and drives being all but level, they are thoroughly available, which is not the case at Funchal in Madeira.

Hamilton, the capital and seat of government, pop. 2000, is pleasantly situated on rising ground overlooking the sea, on the principal island Bermuda, fifteen miles long and two wide. There are good hotels, pleasant English society, and all the comforts and luxuries of English life. The climate, as I have stated, is decidedly insular, mild, with a moist air and a considerable amount of wind and rain. These are the characteristics of all insular climates, with variations dependent on latitude, size, elevation above the sea, and height and direction of mountains.

APPENDIX II.

The Genoese Riviera in Spring : Switzerland and Swiss Mountain Stations in June and July.

Although I have spent nineteen winters at Mentone, I had never before 1878 remained there during May, my annual explorations of the Mediterranean, and the obligations of London practice, having hitherto obliged me to leave in April. My estimate of the climate in May was therefore founded on the reports of others, not on personal experience. A careful study and analysis of the temperature and meteorological conditions existing at Mentone in May and early June of this year, combined with previous experience of the Mediterranean generally in May, have led to the following conclusions. The fear of extreme heat in May in this, as in all other regions of the Mediterranean, which leads winter health-emigrants from the north to make a precipitate return north in April, or at latest early in May, is not substantiated by facts. This fear is founded on the often summer warmth of February, March, and April, but ignores the systemic cold north-easterly winds which usually prevail in Europe in April and May, and which are felt all over the Mediterranean.

This last May, the nightly minimum temperature at Mentone varied between 56° and 60° Fah. It was never once above 60°. The daily maximum varied between 60° and 70°, and was never above. These data correspond to the calculations of M. Brea, which give 63° as the medium temperature of the month. The heat was not at all unpleasant, being merely that of midsummer in the south of England, whilst the vegetation (trees, shrubs, and flowers) was that of July in southern England. We were glad to

look for shade in midday, to don summer clothing, and felt
as we feel in England in a fine warm July. The wind was
generally south-west, and clouds formed every afternoon on
our higher mountains (four thousand feet), but there were
only a few showers, and the rain-laden clouds passed over us
to discharge their moisture in Switzerland, where it rained
nearly all March, April, and May. The weather on the
whole was most enjoyable for all who were well or merely
ailing, as is midsummer with us.

At the same time, I think that chest-invalids and all
whose case requires rather a cool bracing atmosphere than
a warm one, do well to leave the Mediterranean by the end
of the first week in May, say for the Italian lakes or Geneva,
where warm weather comes usually a month or six weeks
later. With them the question is not whether they can
stand midsummer heat without actual injury, but what
climate and what locality is likely to do them the most
good. They can scarcely, with safety, return home to the
north until the end of May, and a cooler climate in May
than that of the shores and islands of the Mediterranean is
likely to brace and benefit them.

To those who are not so situated, who would not fear
midsummer weather in southern England, who would not
try to escape from it by going to our north-east coast, or
to the Highlands of Scotland, May is a most enjoyable
month at Mentone, or elsewhere in the Riviera, and in the
Mediterranean generally. The air is pleasantly warm only,
pure, and balmy, and vegetation luxurious; the sea is warm
enough for sea-bathing (65° to 68°), and merely a passing
shower to be expected, just enough to lay the dust and to
water the flowers. Such, however, I have found to be the
state of the climate in May all over the Mediterranean, from
Gibraltar to Constantinople, in Corsica, Sardinia, Corfu,
Malta, Algeria, Tunis, Asia Minor. I have seldom actually
suffered from heat, and have only found the thermometer
occasionally above 70° or 80° in the day, and 60° or 65° at
night. I do not speak, of course, of the temperature in the

South Mediterranean when a sirocco or south-east desert-wind reigns. In Algeria, for instance, I have known the temperature on the border of the desert in April, at 96° Fah. for forty-eight hours. But the desert wind thus raises the thermometer at any time in spring on the south shores, or in the southern regions of the Mediterranean, for a few days, and this fact is one of the drawbacks of the south Mediterranean as a spring-health residence.

With June, however, the heat became more intense and ceased to be agreeable. M. Brea's medium for that month is 70°, deduced from a day maximum of 74° to 76°, and a night minimum of 64° to 66°. When this is the case, the nights begin to be oppressive, the days are unpleasantly warm, and it is well for all who can go to a cooler region to do so, even the sound; as to the sick and weak, it is folly for them to remain if they can possibly get away. A medium temperature of 70°, continued for weeks and months, develops in the human organisation the susceptibility to all tropical diseases, and the Mentone medium for July and August is five degrees more, 75°. That such is really the case is apparent from the medical literature of our military stations in the Mediterranean, which abounds with the description of serious disease, febrile, malarious, dysenteric, and other, developed by such a temperature. Mentone may be assimilated to the warmest and most sheltered regions of the Mediterranean, owing to its mountain protection. A large population placed under the same hygienic conditions as the population of Malaga, or as our troops at Gibraltar, Corfu, or Malta, would, no doubt, develop the same climatic diseases under the influence of a temperature for three months presenting the medium of 70° and 75°.

In conclusion, I would say that my experience this spring corroborates that of my colleagues at the sanitaria of the Riviera. Invalids had better not leave the locality where they have wintered, or its vicinity—Hyères, Cannes, Nice, Mentone, San Remo—until the end of April, on account

of the cold east winds of the north. This rule, however,
does not apply if they are well enough, and enterprising
enough, to go further south, as, for instance, to Ajaccio,
Algiers, Tunis, Palermo, Malta. By the end of the first
week in May, however, they had better seek cooler quarters
more north. Those who are well, or quite convalescent,
can remain, generally with perfect comfort and safety, until
the end of May, but not later. Even if the summer is passed
without heat disease on the Riviera, an unfavourable result
is generally felt during the subsequent winter.

SWITZERLAND.

Many of those who winter on the Riviera for health
motives have been told at home, or are told in the south,
that they must spend two or three consecutive winters
abroad. Under such circumstances, as some have broken
up their establishments at home, and others dread the heavy
expense of the double journey, or wish to travel, the desire
is frequently expressed to remain on the Continent in some
mountain sanitarium, where the heat of summer would be
avoided. The mountains of Switzerland are crowded with
such sanitaria, to which the town Swiss, the Germans, resort
in summer as we do to the seaside, not only on health
grounds, but for a change—for a holiday. As many of
these mountain hotels or pensions (boarding-houses) are
very cheap (from 4s. 6d. to 6s. 6d. a day), the object of
economy is attained—a most vital one to many.

Having been obliged hitherto, as already stated, to remain
in England in summer, I had no personal knowledge of
these health-resorts, but have been in the habit for years of
sending patients to several at the head of the lake of Geneva,
which my Geneva medical friends told me were amongst the
best in Switzerland—viz., Les Avants, Glion, Champery,
Morgins, Villars, Les Diablerets, La Rossinière, and Cor-
boyier. I determined, therefore, on leaving Mentone this
year, to examine them carefully, and came direct to Geneva,
arriving June 16th.

GENEVA.

At Geneva (1230 feet above the sea level) I found the change in temperature most agreeable, the thermometer in the day being about 68°; at night, about 58°. But it was raining every day more or less, and I was told that it had been raining for three months, ever since March, which had much damaged the spring season. So the south-westerly clouds which had passed over our heads on the north Mediterranean shore, on the other side of the Alps, without giving us any rain, had broken on the Swiss mountains, and deluged the land. I remained some days at Geneva discussing climate questions with several of my friends, professors of the Faculty of Medicine, and intelligent non-medical Swiss residents. I also had a long conversation with Dr. Lombard, the author of the well-known work, "Le Climat des Montagnes." He was one of the first to draw the attention of the profession to the treatment of disease. by mountain air, and is still a vigorous, hale, intellectual man, although in his seventy-sixth year. He has just published the first two volumes of an important work on Meteorology. It is encouraging to his juniors, even to sexagenarians, to see so aged a man still full of life, energy, and scientific ardour, taking a vivid interest in all the scientific questions of the day.

From the analysis of these various sources of information, I came to the conclusion that, by the time the Genoese Riviera becomes too warm for chest-invalids to remain there prudently, that is, about the end of the first week in May, Geneva and its lake have become a safe residence for them. It must be remembered that those who have wintered in the south are more susceptible to unfavourable spring influences than those who have wintered in the north. There are plenty of good hotels at Geneva, of all classes, and the town is open, clean, and wholesome, so that it may be made a spring residence by those who wish to have the resources of a town, of an intellectual centre, at their disposal. The shores of the lake are dotted with handsome hotels and comfortable pensions, and there are steamboats daily, at all hours, to every part of the lake.

I was particularly struck with the beauty of Evian, less crowded than most of the other health-resorts. At the head of the lake are Clarens and Montreux, one a continuation of the other. They are crowded with hotels and pensions, and, in winter, are full of invalids, principally German consumptives. Montreux has become a kind of Torquay to Central and Eastern Europe. It is very sheltered by the mountains from the north, and fully exposed to the sun. No doubt, for Northern and Central Europe and for Switzerland, it is an exceptionally favoured spot, although not to be compared for a moment with the Mediterranean winter sanitaria, unless a cold climate is what is required. All the houses have huge stoves, and the medical instructions published are very particular as to a proper temperature being constantly kept up inside the houses. So, after all, the climate in which the invalids live is not that of the locality in which they reside, but an artificial one created by stoves; it is simply a stove climate. This remark applies still more forcibly to all cold winter sanitaria, St. Moritz, Davos in Switzerland; St. Paul, Denver, in the United States.

On the Genoese Riviera, in well sheltered situations exposed to the sun, even invalids may live with their windows prudently open, day and night, with scarcely half a dozen exceptions, during the six months of winter.

Clarens and Montreux are more sheltered and warmer than Geneva, and spring refugees from the south may settle there until it gets warm enough to return north, towards the end of May; or, if they wish to remain in Switzerland, until it becomes too warm for the lake, and the mountains have to be sought. At Clarens, there are two boarding hotels that I can recommend from personal knowledge as all that can be desired—the Pension Ketterer-Monard and the Pension Gabarel. I was some days at one, then at the other, and found them very comfortable and reasonable.   The terms are from six to seven or eight francs a day.   By June 23rd, however, I found it becoming close and oppressive, which some years it does much earlier, and the rain appeared to have ceased, so I determined to commence my mountain explorations. The thermometer had been above 70° several days when I left to examine, in succession, eight mountain stations in the vicinity of the lake, generally spending a night at each.   They are the stations with which the inhabitants of Geneva, medical and non-medical, are the best acquainted, and constitute their favourite refuges from summer heat.   They are as follows :—

	Feet.
Lake of Geneva above sea	1230
Glion	2254
Les Avants	3212
Corbeyier	3235
Champery	3389
Morgins	4500
Villard	4003
Les Diablerets	3815
La Rossinière	3000

Glion (2254) is immediately above Montreux, on a picturesque, rocky eminence, and is reached by a tolerably good carriage road.   As the lake of Geneva is itself 1230 feet above the sea level, the hotels, of which there are several good ones, are 1024 feet higher.   This elevation insures more air, a better view of the grand Savoy Alps on the other side of the lake, and makes it a very pleasant residence in early summer and in autumn, cooler than the lake level.   It does not, however, secure immunity from the heats of summer.   Glion can only be considered a step in the ladder of mountain ascension.

Les Avants (3212) is a thousand feet higher, one hour's drive, with a good hotel.   I found it decidedly cooler than at Glion.   Both ascending and descending we passed through one of the dense, cold, damp fogs or clouds, which I have described, hanging on the mountain side.   I was

R

told that the nights are always cool, but that the days are warm in July and August.   It is a favourite summer resort with the English residing at Geneva.

All the other places mentioned are in the mountains that border the Rhone valley, at the upper extremity of the Lake of Geneva. They are all within a few hours' drive of Aigle, the second railway station above Villeneuve.   Had time not been an object I should have settled down at the Hôtel des Bains, a large and commodious house, a mile and a half from the station, in a most lovely situation, at the mouth of a picturesque Fir-clothed Swiss gorge or ravine, with walks winding along a wide brawling torrent (La Grande Eau), and in the Fir forest, in every direction. From this hotel, as a centre, the mountain stations I have named can all be reached in a few hours, between breakfast and dinner, the ascents taking from two to four hours, according to elevation and distance, the descents about half the time.   In this way, if the weather is not too hot at Aigle, they may all be visited in succession before a final decision as to residence is made.   Another plan is to take a carriage by the day, as I did, from the livery stables opposite the railway station, and to drive to each mountain station in succession, spending one or two nights at each.   There are stages from Aigle to most of them in summer; the price of the carriages is tariffed.   I paid fifteen francs a day for a one-horse carriage, with a gratuity to the driver.

Corbeyier (3235) is just above Aigle, to the N.E., and is reached by a pretty sheltered road which passes through Yvonne, where the well-known white wine is made.   The hotel is a comfortable mountain pension, rather humble in character and cheap (five to seven francs per day), with a beautiful view of the valley of the Rhone, and of its grand mountain scenery.

The situation must be exceptionally sheltered, for the vegetation was still luxuriant.   The trees were principally Beech and did not at all present the Alpine character; there were Walnut-trees and Cherry-trees in the immediate

vicinity. Corbeyier would do for June, and until too warm
in July, temperature 66° day, 50° evening (June 29th).

Champery (3389) is on the other or S.W. side of the
Rhone, at the upper part of a wide, open fertile valley called
Illiez, which narrows as the higher mountains are neared.
The village is immediately opposite the mountain and
glacier of the Dent du Midi (10,450). The view is glorious,
and I was told that the vicinity of the glacier imparts a
pleasing freshness to the air, but I found it still rather
warm and close in the day. The thermometer was 68° at
5 P.M. 64° at 10 P.M. The scenery around is thoroughly
Alpine. The trees, as we ascended the Illiez valley, were
spruce Firs, nearly the only Conifer seen in this part of
Switzerland, Linden, Ash, Walnut, Poplar, Beech, Cherry,
Apple, Pear, and Elderberry. At Champery there were only
Beech, Firs, and Cherry. There were some of our June
summer flowers in the garden, and grass was being cut
everywhere. At this elevation (3300), and a little above,
and in this part of Switzerland, I found the same vegetation everywhere. I came to the conclusion that Champery,
like Corbeyier, would do very well for June and July, but
that in the great heats of summer it would be advisable to
ascend to a higher elevation. Hotel pension rustic but
comfortable.

Morgins (4500) is situated at the head of a side valley
which branches out on the W. from that of Illiez, at Trois
Torrents. This latter village is also a health station, with a
comfortable rustic pension. Morgins is above a thousand
feet higher than Champery, and the greater elevation made
a marked difference in the climate. All tree vegetation,
except the spruce Fir, had disappeared. The Firs, however,
formed dense forests behind the comfortable pension hotel
built of wood, clothing the sides of the mountain amphitheatre which rises behind, at the head of the valley.
Through the forest rushed and foamed a wide mountain
torrent from the glacier-clad mountains in the distance.
No glaciers were visible; nothing but Fir forests meeting

the eye on nearly all sides. The local authorities and the residents expressed the opinion that it is a great advantage that the air direct from the glaciers should not be felt. I incline myself to think that they are right.

The thermometer was at 61° when I arrived at 2 in the afternoon, and went down to 56° at 8, to 54 at 10 P.M. I was glad to have an additional blanket. I was told by persons who were in the habit of passing their summer there, that the night temperature remained cool during the warmest months, July and August, and that even if warm at mid-day with a fierce sun, it was always pleasant under the trees in the forest. This forest is one of the finest I have seen in Switzerland, reaching nearly to the door of the hotel. There is a lovely gently ascending walk, a mile or two in length, along the bank of the torrent quite in the shade. The ground along the bank was covered with Alpine and lowland spring flowers. The proximity of this fine Fir forest to the hotel is a great advantage, for the sun being very high in the heavens in midsummer in this latitude its mid-day rays are scorchingly hot at any elevation, even when the air itself is cool. At Morgins visitors can pass the day in the forest and thus avoid the sun's rays, whereas in other stations, quite as high or higher, they are driven indoors for want of forest shade.

There is at Morgins a strong chalybeate spring, a few hundred yards from the hotel, which attracts many invalids, as they thus combine cool mountain air with an iron water treatment.

Taking everything into consideration, I think Morgins is the most desirable high elevation station that I have yet seen in Switzerland, and I question whether there is any advantage to be attained by ascending higher. Even here, in fine weather, I had to pass through a dense layer of cloud or damp fog in descending. My Swiss friends questioned whether it was not too high for chest cases. For consumptives, they said they scarcely dared send them so high, fearing the dry, keen mountain air, would precipitate the progress of disease. For persons who are well, and able to climb mountains, it is an admirable centre for excursions. This really valuable station is but little known or frequented by the English. At the Champery and Morgins hotels the board is from 6 to 7 francs.

I was sorry I could not stay longer at Morgins, to explore the neighbourhood, which seems full of mountain interest, but I had many places to visit, so I descended the Illiez valley rapidly to Monthey, crossed the Rhone, reached Aigle, and the same day ascended to Villard.

Villard (4003) Grand Muverain and Bellevue Hotels.— This is a favourite station with the Genevese, and with the English acquainted with these regions. It is reached by a good winding road from Aigle in 3½ hours, through a fertile country, dotted with fruit-trees—Apple, Pear, Cherry, and Walnut; higher up come Beech, and then Fir-trees. The hotels are comfortable, well-built, first-class houses. The view of the valley of the Rhone, of the mountains on the other side of it, of the Dent du Midi and Mont Blanc, and of the mountains around, Grand Muverain and Diablerets, is magnificent. The storms here—I witnessed one—are most imposing, as also the effects of cloud, rain, and mist filling the valley of the Rhone, and rolling over and on the surrounding mountains, with blinding flashes of lightning and peals of thunder. There are Fir forests within easy reach, but none so near as at Morgins. The accommodation, however, is very superior, an important feature to those who depend on hotel comforts. The temperature was cool in the day, 62° (June 29), cold at night, 52°. The vegetation was that of the north of Scotland—Laburnums and Lilacs in flower. I should think the summer heats might be passed here very pleasantly, by avoiding the glare of the sun, and keeping indoors when mists and fogs reigned.

The Hôtel des *Diablerets* (3815) is situated at the village of *Les Plans,* in the upper part of the Ormont valley, at the outlet of which is the Hôtel des Bains, Aigle. When

at Villard it is necessary to return to Aigle to ascend by
the Ormont valleys, although the two localities are only
separated from each other by a high mountain, the *Chamos-
saire* (6953). The ascent by the Ormont valleys, upper and
lower, occupies four hours. The gorge is deep and picturesque, with steep banks clothed with Fir-trees, and with
a wide torrent rushing over the rocks at the bottom. The
road winds along the sides of the mountain, often on ter-
races cut through the rock. The Hôtel des Diablerets
seemed in the same conditions of atmosphere and temperature as those at Villard. The air was pure, cool, pleasant,
60° in the day, 51° after sunset. The vegetation was Alpine
just above the fruit-tree zone—Firs, with a few Cherry-
trees. The hotel is within an hour's walk, through a Fir
forest, of a rocky amphitheatre at the base of the widespread glaciers of the Diablerets (10,043), over the sides
of which pour a number of cascades. They unite and
form the river torrent—the Grande Eau—which occupies
the gorge we had ascended. The hotel is a comfortable one,
but not so luxurious as the Grand Muverain at Villard.

La Rossinière (3000) is reached in four hours from Les
Diablerets, six from Aigle. The Pension Henchoy is a
quaint, large Swiss chalet, the largest, it is said, in Switzerland, built entirely of wood, and covered with carved
inscriptions. The village and hotel are situated at the
bottom of a most picturesque Alpine valley, the mountain
sides of which are clad with Fir and Beech-trees, and clothed
with green pasture. La Rossinière is on the eastern side
of the Etivas mountain ridge, which divides the watershed
of the Rhone (Mediterranean) from that of the Rhine
(German Ocean). The river in this valley runs east. Here
I found the same thermometrical conditions as at Champery
and Corbeyier, at nearly the same altitude. A rather
warm day temperature, 66° to 68°, and a cool night
temperature, with our summer June vegetation. In the
woods and pasturages the flowers were innumerable; they
were everywhere cutting the grass, as elsewhere at this
elevation.

From the analysis of what precedes it will be seen that I examined, during the last fortnight in June, four mountain stations, at an elevation of from 3000 to 3400 feet—Les Avants, Corbeyier, Champery, La Rossinière—and found in all the temperature and vegetation of the same date in Kent and Surrey, in an average year. I examined three stations at an elevation of from 3800 to 4500 feet—Les Diablerets, Villard, Morgins—and met with the temperature and vegetation of Northumberland or of the Highlands of Scotland at that time. Of these three Villard presented by far the greatest amount of comfort, and Morgins seems to present in the highest degree the requirements of a high mountain station for health purposes—temperature, air, forest, chalybeate water. All these higher stations are points of departure for excursions into the higher mountains, to the peaks and the glaciers, which must greatly increase their charm to the strong.

Being aware that there are several esteemed mountain stations in the vicinity of the Lake of the Four Cantons (Lucerne), and also of the lakes of Brienz and Thun, I determined to extend my explorations to these regions, so on the 2nd of July left La Rossinière for Fribourg and Lucerne, nearly two thousand feet lower down. This journey gave me an illustration of the uncertainty of mountain climates, and of the dangers to which those who suffer from chest disease are exposed, even in mid-summer.

It began to rain soon after I left La Rossinière, and rained for thirty-six hours, until the morning after reaching Lucerne, July 4th. The rain prevented my visiting *Le Lac Novi*, a station well spoken of near Fribourg. All the way the higher mountains were enveloped in cloud, and invisible. So it was when I looked out early at Lucerne, although the rain had ceased. By ten o'clock these clouds had melted, and then I saw all the mountains that surround the Lake of the Four Cantons covered with a deep coating of snow. The Righi had a layer of snow several inches deep, some distance below the hotels, apparently half-way down. Mount

Pilate was half-covered with snow. It was very cold—
58° only at mid-day, and all day. All the mountain pensions above 3000 feet high, I subsequently heard, were
enveloped in snow. I may judge of the effects of such an
outburst on invalids by its effect on myself, who am now
quite well for a man of my age (62). It suddenly checked
the biliary secretion of the liver, and brought on congestion
of that organ and a bilious attack, whilst the upper air-
passages became irritable and slightly inflamed. It took
me several days' nursing and doctoring in the comfortable
Schweizerhof Hotel to recover.

LUCERNE AND ITS MOUNTAIN SANITARIA.

The cold chilly weather only lasted forty-eight hours,
and after a few days' quiet I again resumed my mountain
explorations.

Lucerne (1437 feet above sea), like Geneva, has an old
town and a new one. The latter is principally composed of
fine monumental hotels, built on the shores of the lake,
which they overlook. The entire town, old and new, is
exceptionally clean and bright. The mountain scenery
around is glorious, magnificent, and it may safely be made
a long resting-place by those who prefer cities to the country,
weather permitting. The fine luxurious hotels are rather
expensive, but there are numerous cheap boarding-houses
(pensions) in and around the town.

The lake assumes the form of a Maltese cross formed by
four unequal arms, E. W. N. and S., with a prolongation
from Brunnen due south. Steamers start every morning
from the quay at 10 A.M. for every part of the lake, returning
in time for the 5.30 and 7.30 hotel dinners. These steamers
afford an easy and pleasant way of exploring every region of
the lake, and every place within a few hours' drive of it,
without sleeping out. The plan is to take the 10 A.M.
steamer to the foot of the road that leads to the sanitarium to
be visited, there to hire a carriage and to lunch at the
hotel, descending in time for the return boat.

I ascertained that the principal mountain sanitaria within easy reach were:—

	Feet.
Axenstein	2300
Axenfels	2200
Kurhaus Sonnenberg	2772
Burgenstock	2854
Engelberg	3314
Righi Culm and Righi Hotels	3000 to 5906

The *Kurhaus Axenstein* (2300) is a large, elegantly appointed hotel, in extensive well-wooded grounds, situated on the mountain side, above the town of Brunnen, about two hours by steamer from Lucerne. There is a grand view of the mountains that bound the other or south side of the ·lake. *Axenfels* is also a large comfortable hotel, a hundred feet lower down, with an equally splendid view, but the grounds are less extensive and wooded, and it is rather cheaper. Both these hotels form delightful mountain residences for June and July, until the great heat of summer renders it desirable to ascend higher. There is a good carriage road from Brunnen, by which the ascent occupies an hour only. The season is said to begin as early as the 15th May.

The *Kurhaus Sonnenberg* (2772) is a very large hotel, with above three hundred beds, on the opposite or south side of the lake, and about six hundred feet higher than Axenstein. It is also sheltered to the north by a higher mountain, at its back (Seelisberg, 6313 feet). It has a large German connexion, as have all the Lucerne mountain sanitaria, and is full to overflowing during June, July, August, and September. The landing place, Treib, is just opposite Brunnen, and carriages, tariffed, are always waiting. It is a little town, with large reception and dining-rooms, a village of accessories and outlying works, and music three times a day in the German fashion, with all the outdoor life and activity of a German watering-place, high up in the mountains. The views are as grand as at Axenstein, but command the mountains that skirt the north side of the lake. It looks

down on Axenstein and Axenfels, and must be a charming
mountain-residence for those who like music and the bustle
and animation of a watering-place.   I found the air cool
and pleasant, and being higher and directed to the north it
is said to remain cool longer than at Axenstein, in hot
weather.

Burgenstock (285·4).   Here there is a most comfortable
and elegant hotel, much smaller than those just described.
It is situated on a saddle-back projection, promontory, or
ridge of the Burgenstock mountain (3721) which advances
into the lake on its south side, opposite the Righi and
nearly opposite Lucerne.   From thence the lights are seen
every evening; they directed me to it.   The grounds are
extensive, well wooded, and the view unsurpassed, extending
far away over mountains and lakes from both sides.   Indeed,
there is not a window in the hotel that does not command
a glorious view.   It is a much more retired, quiet, refined
place than Sonnenberg, and struck me as one of the most
enjoyable residences that I have seen in Switzerland.   I was
told that in early spring it is rather windy from its exposed
position, but by the middle of June the winds have usually
abated, and residence may begin to end merely when the
weather becomes too warm.   The hotel is cheerful and
elegant, and appears well managed.   The grounds and
walks are kept like those of an English gentleman.   It is
but little known to the English, and is principally frequented
by first-class Germans, I was told.   The north-east view
is the same as the one from the Righi, extending over the
lakes to the north-east of Lucerne.   The hotel is easily
reached by steamer from Lucerne to Stansstad in one hour,
with a drive of one other hour.

Engelberg (3314) is a mountain amphitheatre at the
summit of a very picturesque beech- and fir-clad valley,
through which descends impetuously a torrent or river.
It is reached by steamer from Lucerne to Stansstad and
by a road which traverses the plain at the foot of the
Burgenstock and the small town of Stans.   From Stans the

rise for the first seven miles is all but imperceptible, and the valley wide and fertile. The last six miles of the road is through the gorge described, which opens out into a green mountain amphitheatre bounded on three sides by lofty snow-clad mountains, half-covered with forests: the *Titliss* (10,627), the *Great Spannort* (10,515), the *Little Spannort* (10,382). There is quite a large village at Engelberg, and as it is a very favourite station there are several hotels and pensions, but the most comfortable is unquestionably the Hotel Sonnenberg, a large well-managed, recently built hotel. The foot of the glaciers is within an hour or two's walk of the hotel. Engelberg is recommended by German physicians principally as a residential mountain sanitarium for the entire summer. It is said that the glaciers which surround it on three sides, at a higher elevation, keep it cool, even in the warmest weather, although I have always found the days warm at this elevation in Switzerland in warm weather; and the daily breeze or current up the valley is a warm one, coming as it does from the plains.

The Righi, a quadrangular mountain, standing out in the Lake of Lucerne, which bathes two of its sides, is the great central Swiss attraction to tourists, principally on account of the railway to its summit. Here at Righi Culm (5906) there are now two splendid hotels, entirely filled by tourists who sleep there to see the sun rise in the morning. Every afternoon in summer several hundred arrive to leave next morning. It is too exposed a situation, at this elevation, to be a desirable position for invalids, and none remain there. In the middle of summer if it rains lower down there is snow, fog, and mist at Righi Culm. I ascended on the 7th of July to find a cold impenetrable fog, with rain from the Kaltbad (4728) to the summit. Three days before the entire mountain above 4000 feet had been covered with snow. On a former occasion, many years ago, I spent a night there in September, in cold but delightful weather.

At a lower elevation, *Kurhaus Kaltbad* (4728), *Righi Staffel* (5210), *Righi Scheideck* (5407), there are well-built,

well-appointed residential hotels calculated for a mountain
sojourn for those who merely require coolness in hot weather,
with pure bracing air. They are, however, not calculated,
in my opinion, to afford a safe residence to those who suffer
from chest disease, acute or chronic. The days are often
too hot from sun glare in a dry atmosphere, without forest
shade, the nights too cold; the difference between the day
maximum and the night minimum also, is too great. Moreover there are the occasional bursts of bad weather, snow,
fog, cloud, and cold rain, which I witnessed, to be encountered.

LAKES BRIENZ AND THUN—THEIR MOUNTAIN SANITARIA.

My next move was to the Lakes Brienz and Thun, the
sanitaria of which I also wished to explore. I crossed,
therefore, the Brunig Alps by the pass so called, leaving the
Lake of Lucerne at Stansstad, and passing by Alpnack,
Sarnen, and Lungern. At the summit of the Brunig Pass
there is a homely, tolerably sheltered mountain hotel or
pension: Hotel Brunigkulm (3390), which would offer a
picturesque, cheap, healthy, mountain retreat. The stations
I subsequently visited are :—

	Feet.
The Giesbach	2400
Interlaken	1863
Grindelwald	3468
Lauterbrunner	2615
Murren	5348
St. Beatenberg	3767
Weissenberg	2940
Gurnigel	3783

The Brunig Pass road strikes the Lake of Brienz (1857)
at the town of that name. Thence the steamer takes the
traveller in a few minutes to the landing-place for Giesbach.
The road ascends for 500 feet through a Fir forest, to terminate on the terrace of a splendid hotel, recently built
(2400). There is an older building in the background,
which now constitutes the pension or boarding-house, the

palatial hotel not receiving boarders. The chief attraction to Giesbach is its succession of beautiful waterfalls. A constant stream of tourists fill the hotel, remaining one, two, or more days. The level walks in the Fir forests, the picturesque views, the beautiful terrace and garden, and the numerous mountain excursions, combined with the comforts of an elegant well-appointed hotel, and the charm of good music, make Giesbach a delightful place. Invalids, who appreciate luxury even in the mountains, might safely spend the early summer here, merely leaving to ascend to a higher station if the heat becomes too great.

Interlaken (1863), between Lakes Brienz and Thun, is now a good-sized town, with many comfortable first-class hotels. It is a great centre for tourists intending to ascend the neighbouring mountains and glaciers, but scarcely a desirable residence for chest invalids. They had better avoid agglomerations of mankind, and seek the wilderness, with town comforts, now to be met with in every part of Switzerland.

Grindelwald (3468), from its elevation, consequent coolness in early summer, and from the existence of several comfortable hotels, offers conditions favourable to invalids seeking for mountain air; but there is a great objection : the hundreds of tourists going and coming, and the dust and commotion connected therewith. Such being the case, I cannot recommend Grindelwald as a residence for invalids of any kind. It was very hot the day I was there, July 18th; at 6 A.M., 58°; at mid-day, 76°; at 2, 84°.

Lauterbrunner (2615), two' hours' drive from Interlaken, separated from Grindelwald by the Wengernalp, is situated at a rather lower elevation, in a pretty valley, with picturesque walks and comfortable hotels; but it is open to the same objection as Grindelwald. A stream of tourists is constantly arriving and departing, as the mountain pass between the two localities is perhaps the most frequented one in Switzerland. It is also the point of departure for the ascent to Murren.

Murren (5318) is situated on an uneven terrace or ledge three thousand feet above Lauterbrunnen, and exactly opposite the magnificent mass of glaciers and snow-covered mountains formed by the Eiger, the Mönch, the Jungfrau, and the Matterhorn, from which it is separated by the deep Lauterbrunnen valley. There are two really good comfortable hotels, and the locality is becoming a favourite mountain residence. I do not, however, hesitate to say that, although suitable for anæmic, debilitated persons, it is altogether unfitted for chest invalids. The ascent from Lauterbrunnen is all but precipitous, and the descent can only be made on foot, or in a sledge, for the greater part of the distance. The extreme temperatures, as in all the high elevations, are very great. On the 4th of July of this year, 1878, there were three inches of snow on the hotel verandah; on the 19th, the date of my visit, at two o'clock, the temperature was 80° in the shade.

St. Beatenberg (3767) is reached in three hours from Interlaken. The road winds through Fir forests on the side of the mountains which skirt the north side of the Lake of Thun until it reaches the village of St. Beatenberg, a favourite retreat for both Germans and Swiss, sheltered from the north by the higher mountains and open to the south towards the lake. There are several hotels and pensions, but the principal one is the Kurhaus St. Beatenberg, kept by a Swiss doctor, about a mile further on. The views of the lake and of the snow-covered mountains that surround the Jungfrau are entrancing. The altitude is sufficient to secure cool nights even when the days are hot. It is principally recommended to nervous patients. The hotels and pensions are generally full of German and Swiss invalids, so that application must be made beforehand.

Weissenberg (2940) is one of the most celebrated and most frequented watering-places in Switzerland for diseases of the lungs, phthisis, chronic bronchitis, and asthma. It is situated in a narrow, deep Fir-clothed, picturesque gorge or ravine, which opens out into the Simmenthal valley, about three hours' drive from Thun, and the same from Berne. The wild wooded ravine has to be ascended about a mile and a half from the village in the Simmenthal valley before the bath buildings are reached. They are composed of a new *établissement* or Bath Hotel, a comfortable, well-built, and well-appointed house, and the old bath-house ten minutes' walk higher up. This is a more humble building in appearance, and is reserved for the poorer patients, called second and third class. A wild torrent falls in cascades down the bottom of the valley below the hotel, carrying off the whole drainage of the place, and a jet of water nearly a hundred feet high rises in front of it, falling in a widespread spray. The situation realises the ideal of a wood-clothed mountain gorge, and is very picturesque.

The far-famed waters contain a considerable amount of sulphate of calcium, and a smaller amount of sulphate of magnesium, just enough to counteract the astringent effects of the first-named salt. I found the establishment crammed with consumptive and bronchitis patients, above 400, from all parts of Switzerland and Germany. The night I was there twenty slept in the public drawing-room, and I only found refuge in the consulting room of the aimable and learned physician, Dr. Schnyder. The faith that every one in the place had in the curative properties of the mineral water, in these chronic constitutional diseases, struck me with amazement; physician, patients, administrators, all seemed conscientiously to believe that a few tumblersful a day of this mere gypsum impregnated water, for three or four weeks, repeated two, three, or more years, would effect a cure. The poor sufferers all but fought for admission to this modern fountain of Jouvence, crowding the village below. One feature in the medical history of the spa greatly impressed me—viz., that with all it was a matter of creed that the moister the atmosphere was the better it was for the patients! The doctor told me, with exultation, that constantly at sunrise, the hygrometer marked saturation, just the condition, he said, required for chronic lung diseases, such as phthisis and bronchitis. At many places previously visited I had been told, with equal exultation, that the extreme dryness of the air at from 3000 to 5000 feet elevation was the prominent feature in the climate, the one that rendered it the right place for the same class of sufferers! The truth probably lies between the two—I presume lung sufferers find relief in the mild moist air of Weissenburg, or they would not crowd there. Nevertheless, I should prefer a drier atmosphere, a higher elevation in warm weather, and a less crowded place.

Gurnigel (3783) is a well known sulphur bath establishment, about three hours' drive from Thun and from Berne. It is situated at a considerable elevation on the sloping side of the mountain of that name (3783). The hotel built on a very large scale, inasmuch as it affords shelter to between four and five hundred patients, is under the intelligent management of the owners, the Frères Hauser, the proprietors of the Schweizerhof at Lucerne, of the Geisbach, and of several other first-class hotels. They showed me over their vast establishment, always full to overflowing in summer, and I was much struck by their liberal and intelligent direction of this great ship in the mountains, far from all ordinary sources of supply. The grounds and woods are most extensive, covering several square miles of the mountain side, and traversed by good roads and paths in every direction. The views are lovely and open, extending far away over mountain and valley. The waters are valuable, containing, in addition to sulphate and carbonate of lime, sulphuretted hydrogen in considerable abundance. There are two sulphur springs, one mild, the other strong, temperature 52° F., and a chalybeate spring. Thus a sulphur or iron treatment can be carried out at these baths under favourable conditions of coolness from mountain elevation, with all the comforts and luxuries of a first-class hotel. I left Gurnigel with the impression that it is one of the best and most enjoyable mountain residences for invalids that I had found in Switzerland. But day heat is not escaped even at an elevation of 3783 feet. On the 22nd July the heat was 80° F. at 8 P.M., at 10 it was still 78°. Next morning, July 23rd, at 7.30 A.M., it was 74° F. at 2 P.M. 78°.

Probably mountain air, quiet, and rest have as much to do with the patient's improvement in most cases as the mineral water, but at Weissenberg they do not think so. The hygienic adjuncts are considered secondary.

A singular feature in the therapeutics of Phthisis and of other chronic diseases on the Continent, and especially in France and Germany, is the belief in the curative powers of sulphureous mineral waters. Those of the Pyrenees, of Eaux Bonnes, Eaux Chaudes, Bagnères de Luchon, and Cauterets are much relied on. French patients, belonging to the better classes, are all but invariably sent in summer to these or other sulphur springs, under the conscientious belief that taking the waters, a few tablespoonsful or a tumblerful or two daily, for three or four weeks, will arrest a constitutional disease that has lasted months or years, secure the retrogression of all morbid conditions, and lead to the recovery of health! There is not even a trace in our English medical literature respecting Phthisis of this belief, which I am convinced, from clinical observation, is totally unfounded.

These views appear to me to be, in a great measure, the result of the peculiar discipline and organisation of mineral watering-places in France, more especially. One or more medical inspectors are named to each spring by Government, and usually all but monopolise the practice. The medical lives of these physicians are thus spent in administering the waters of their spring to all comers, and by degrees the concentration of thought on the virtues and curative power of the spring, and on the traditions connected therewith, as I have elsewhere stated, warps their judgment. Thence they attribute all the favourable changes that accrue to their patients at the time, or afterwards, to the curative powers of the mineral water they have taken during a few weeks.

S

Nearly every physician at a mineral spring writes a pamphlet or a book on his waters, so that there are tons of printed matter in existence on the mineral springs of France and Germany, and each spring is presented in the light of a curative power of the highest character, particularly for some special diseases. A very remarkable fact is, that none of these writers admit that sulphur, iodine, soda, potash waters, taken at home, medicinally, can produce the same marvellous effects. They must be taken at the spring itself, the "alchemy of nature" giving to them at the source a virtue or a mysterious curative essence, which it is impossible to imitate. Even the waters themselves, bottled and taken on the spot, away from the spring, are said to be immeasurably less curative.

It must be well understood that I am only criticising the exaggerated belief in the powers of these mineral waters. There is great and undeniable advantage to be derived in cutaneous and mucous membrane diseases and in rheumatic affections from sulphur waters, in gout, liver, and kidney diseases from alkaline waters, as also in the treatment of many other forms of chronic disease. Indeed, whilst continental practitioners exaggerate the power and virtues of mineral waters, most British practitioners all but entirely ignore their real merits and the great assistance they are capable of giving in chronic maladies.

One reason of the great reliance of those who inhabit central Europe, in mineral waters, from the emperor to the peasant, is that it has become the habit, the fashion, for the annual holiday to be taken at some spring or other, just as we take it at the seaside or in the Highlands of Scotland. The universal desire is to combine profit with pleasure; and to go to some spring that will, as if by magic, cure disease and renew life. The wishes and desires of the public chime in with the blind confidence of the spring physicians in the virtues of their waters, and thus the public and the faculty harmoniously join in fostering a therapeutical delusion, to their mutual benefit. The patients, however, in addition to the mineral waters in which they place such implicit reliance, obtain leisure, mountain or country air, and brain rest.

With Gurnigel I concluded my examination of the mountain sanitaria of the Geneva, Lucerne, Brienz, and Thun lake districts. The convictions, based on former summer travels in mountain regions, and contained in the body of this work, were confirmed. After careful exploration and study of the mountain sanitaria of this part of Switzerland, I fail to see any meteorological or climatic conditions which impart any peculiar virtue to them in the treatment of chronic diseases of the respiratory organs. On the sea-coast, and on the moors, downs, and plains of the British Isles, on the shores of Northern France, of Belgium, and of Holland, the same conditions of medium temperature and of purity of air are obtainable, without the drawbacks pertaining to the Swiss mountains. These drawbacks are, the distance, the fatigue of travelling, the heat of the plains, the changeabilty of the weather in the higher regions, and the extreme variations of the thermometer from heat to cold and *vice versâ*, and of the hygrometer from dryness to moisture. As regards invalids generally, the higher mountain elevations in Europe may be beneficial for a short residence, but the deductions recently published by Dr. Lombard are against the desirability of prolonged residence. At page 287 of the first volume of his recent work on " Medical Climatology" he says :—

" The facts which we have quoted on the development of the human stature show that, with certain exceptions, the stature becomes lower as the higher elevations of the Alps and Pyrenees are attained ; as also in Savoy and Piedmont, where, without exception, the diminution of stature was in direct ratio to altitude.

" As to the infirmities other than stature which may exempt from military service, we have seen that they increase with altitude in the Pyrenees, as also in the French Alps, and that this result was more especially striking on comparing different parts of Savoy and Piedmont, where the number of exemptions increases in direct ratio to the altitude. We have also seen that if it was difficult to establish, with some probability, the comparative duration of life in the plains and in the mountains, it was, on the contrary, very positive that mortality increases with the elevation of the soil. Is it the same in the higher regions of the temperate zone? This is what we cannot affirm, in the absence of precise documents. But what appears infinitely probable is that the weakening of life recognised for medium elevations must be the more pronounced the higher the altitude."

There is a class of sufferers to whom I feel very sympathetic, and to whom I cordially recommend the mountain stations that I have described. It is my overworked, over-strained fellow-practitioners. A few weeks spent at any of the mountain stations I have described, wandering among the Fir forests, the mountain peaks, and the glaciers, would do them a world of good, and be a source of intense satisfaction. Every one of these Alpine hotels and sanitaria is surrounded by Alpine walks, excursions, and ascensions, which constitute the principal charm of residence to those who are well enough and strong enough to undertake them. Any of the lake districts I have described—Geneva, Lucerne, Brienz, and Thun—would be quite enough for one tour, the more so as in each there are many interesting places to visit besides those I have described, should there be the will and the leisure.

In my opinion the mountain regions of Switzerland, and of central Europe in general, are more suited to this class of visitors, to town people overworked but physically well, and to athletic young men seeking for kingdoms to conquer in the glaciers of the high Alps, than to invalids of any description. I always, however, except the inhabitants of the plains and cities of these regions, and those of central Europe, for they have no other resource to escape heat, and to attain purity of atmosphere in mid-summer, but to ascend mountains if they cannot get to the shores of the North Seas.

It is a general remark that travelling in Switzerland has become very much more expensive than it used to be. To a certain extent this statement is correct, but it is not entirely so. Palatial luxurious hotels have been built at vast expense all over Switzerland. The British and American visitors crowd to them, find them expensive and complain. But the old-fashioned, rustic, wood-built, cheap inns and pensions still exist, and are still frequented by crowds of continentals. Side by side with the large hotels there are generally several small and cheap ones, to which those who wish to limit expenditure should go. They are nearly all noted in Baedcker's useful guide.

APPENDIX III.

The Balearic Islands—Majorca.

On former visits to Spain I had heard that the Balearic Islands were greatly esteemed by the Spaniards as a winter residence for Phthisis, more so, indeed, than Malaga and the cast coast. My wish to visit them was at last gratified. On April 30th, 1877, I left Mentone, crossed the South of France and the Pyrenees, reached Barcelona and took the weekly steamer. Starting at 4 P.M., we arrived at 8 A.M. at Palma, the capital of Majorca, an imposing fortified city (population 50,000) built on an eminence towards the sea.

Majorca is the largest of the Balearic Islands, as its name implies. It has an irregular bay-indented coast, the north side running from north-east to south-west; the greatest width is 45 miles, the greatest length 60 miles. The distance from the nearest point of the mainland of Spain is 120 miles.

Along the northern coast, from north-east to south-west, runs a ridge of mountains, principally composed of limestone, which rises to a height of from 1000 to 5000 feet, with a depth of from ten to fifteen miles. These mountains protect the interior of the island to a considerable extent from north-west winds, but it is entirely open and exposed to the north-east. At the east extremity of the islands, north, there is another block of mountains, which cannot be of much use as a protection from the north-east or the south-east winds, but must serve to precipitate rain in abundance from the moisture-bringing sirocco, or south-east wind, after its passage across the Mediterranean Sea. Between these two ranges of hills there is a vast fertile plain, nearly on a dead level, but slightly elevated above the sea (lat. 38° 38′—40° 6′).

Such are the elements of climate at Majorca. It is an island situated in the west extremity of the Mediterranean, with two ranges of mountains, not very high, and not very

deep; the one running from east to west north, the other
a block of mountains east, enclosing an interior level area,
or plain, with all the characteristics of a limestone soil. In
other words, the soil is adapted to grain crops of all kinds,
to fruits, and to vine cultivation. The mountain ranges
insure a heavy fall of rain in autumn and spring. In
autumn rain falls when the north winter winds set in, and,
laden with the moisture collected in passing over the sea,
impinge on the cooler mountain elevations; in spring, when
the warm moisture-laden winds from the south-east and the
south-west replace the north winter winds, and bring summer
weather with them. In summer there is little or no rain,
for the mountains become too hot to precipitate the moisture
contained in the winds, either north or south.

The power of the sun being great, from latitude, the heat
is intense during the summer months, notwithstanding a
gentle sea breeze which extends far into the interior.
During the period I was in this island, May, the sun being
all but perpendicular, its rays were unbearable without the
protection of a sunshade. The temperature was generally
about 75° in the day, from 66° to 70° at night. The above
meteorological elements contain all that is required to make
Majorca an exceptionally favoured specimen of the Mediterranean insular climate, both as regards agricultural, horticultural, and botanical products. They may be resumed in
a few words—abundant autumn and spring rains, moderate
protection from northern winds, intense heat and dryness in
summer from April until October, with, no doubt, heavy
dews at night owing to a marine atmosphere. But they do
not make it an exceptionally good winter station.

Majorca is thickly populated, its central plain being
studded by villages and towns, the inhabitants of which are
maintained by the product of the land. Every inch of the
plain is cultivated. It is divided into fields, or separate
properties, large and small, by thick stone walls some three
feet high, made of the limestones collected out of the fields in
the process of cultivation. The census of 1860 gives 209,032.

Fruit Trees.—The tree products of the soil are Olive,

Carouba, Fig, Almond, and Apricot trees. They are planted at twenty, thirty, or forty feet distance one from the other, so that the sun reaches the soil around and under them. The soil is also everywhere cultivated with Cereals—bearded Wheat and Barley, occasionally Oats, with Potatoes and other green crops, principally Beans. I saw but few Vines, but was told that they exist in numbers, and that a good deal of wine is made in some districts.

The system of cultivation, I was told, is to grow Cereals one year, the next green crops, the third the ground is allowed to remain fallow. Sheep or cows are pastured on the natural grasses during the fallow winter, the land being broken up by the plough in the spring previous to re-cultivation with cereals in autumn. This is the system followed all over the Mediterranean area. The Cereal crops seemed to me generally poor and meagre, both as to height and ears, but the general product must be considerable, as every crevice and cranny is cultivated, even holes amongst rocks bursting out of the soil.

The fruit trees seem to constitute a very important part of the produce of the land in every region. The Olives are innumerable, and more remarkable for their extreme age than for their size. We have larger Olive trees on the Genoese Riviera, but they can hardly vie with the Majorca ones in the outer evidence of age. They seem to be as old as the Flood, and are rent, twisted, divided, broken, torn in every conceivable and inconceivable shape, representing demons, monsters, dragons, and reptiles. They have often only one, two, or three healthy young branches, like young trees, growing out of a hideous mass or wreck of misshapen roots and timber, the remains of a noble tree one or two thousand years ago. These young, fair branches reminded me of old men's children—fair to look at, but with no hold on life. The produce of oil is immense, but it is carelessly prepared, and is anything but savoury to strangers.

The Carouba or locust trees are fine, vigorous, large trees, as large as fifty or eighty year Oaks, covered with an abun-

dant dark green foliage; many are very old, and twisted like
the Olives. Their presence in such vigour is another evidence
of intense heat and dryness in summer, as they are only
found growing luxuriantly in the hottest and driest parts of
the Mediterranean, such as Syria and the east coast of
Spain. They produce abundantly a large bean, in a pod
which grows directly on the trunk and branches, in a very
odd manner. This bean is most valuable for feeding horses
and cattle in general. Carouba-trees are very numerous in
Majorca, interspersed among the Olive trees, and are grown
for their produce, not for ornament, which is little if at all
thought of in the Mediterranean.

The Fig-trees are also very large and very numerous.
They are often quite timber trees, and as large as those
which I saw at Aidin, near Ephesus, in Asia Minor, from
whence are gathered the figs we receive from Smyrna.
These figs must be an important article of diet as well as of
export. One large Fig-tree, with its double crop and abundant produce, suffices for the fruit wants of a large family.
A few figs, or any fruit, with a piece of bread, make a luxurious meal for the poorer inhabitants of Southern Europe.

The Apricot-trees are large, beautiful timber trees, often
as large as forty or fifty year Oaks, covered with dense
lovely green foliage, and produce fruit in profusion. It was
just beginning to ripen, but was small and flavourless. The
Apricot is clearly suited to a very warm and dry climate.
I found it, on former occasions, growing as a timber tree in
Murcia, in the driest and sultriest part of Spain, and also
in Algeria, near the desert. In these regions it grows to a
size and to a beauty of which we have no idea. No wonder
that its branches constantly die off in our hyperborean
climate; the wonder is that it lives at all in the open air.
These trees, however, are not very numerous in Majorca,
and are generally planted near farmhouses and villages, no
doubt for the personal use of the inhabitants. As the fruit
does not keep they can have little or no trade value.

The Almond-trees are also very numerous, and almonds

must form an important item of export. They are principally grown in groves or orchards in some localities. They
form large, often fine trees, although their habit of dividing
into two or three large branches near the ground is not
conducive to tree beauty.

Cereals grow and flourish in the sultriest climates, merely
because their growth is over and the harvest is garnered
before the sultry weather begins. Thus on May 12 I found
that the cutting of barley was becoming general in Majorca,
and it was evident that in a week or two the "bearded
Wheat" would be ready for the sickle, which, by-the-by, is
the only harvesting implement used. A few ears of Barley
are held in the left hand, and cut with the right. In the
Mediterranean basin the spring rains generally end early in
April, but by that time the Cereals have their roots deep in
the ground, and can conclude their vegetative career without
any further supply of rain. The grain is ready for harvesting by the middle, the end of May, or by the first week in
June at the latest, by which time, only, the very hot weather
of the South begins. Thus it is that bread is always good
in these regions of Europe, the grain crops never being
spoilt by untimely rain, as in the North. Thus it is that
Cereals can be grown successfully in the hottest parts of the
earth, as in the oases of the desert of Sahara. Thus it is,
also, that in dry summers in England, when pastures fail,
there is generally a good grain crop.

Majorca Orange Culture, the Soller Valley.—All over the
western Mediterranean splendid large oranges are sold as
Majorca oranges. They appear earlier than those grown on
the Genoese Riviera or at Valentia and other parts of Spain,
and are more expensive. One of the objects of interest in
my excursion to Majorca was to investigate the orange
groves, and I expected to find the island stocked with them.
To my great surprise, after passing several days in travel
—after crossing it in various directions, from north to south,
and from east to west—I had scarcely seen an Orange tree.
I inquired, therefore, whence all the Majorca oranges could

possibly come from, and was told that they all came from one region, the valley of Soller, situated in the north mountain range, about twenty miles from Palma. To Soller, therefore, I determined to go, taking a carriage, as the road is good.

For the first twelve miles, travelling due north, we merely passed through a portion of the central plain already described. We then reached the base of the northern mountain chain, ascended a series of zig-zags, and crossed the ridge at an elevation of 1500 feet. On looking down at our feet we saw a marvellously beautiful amphitheatre, formed by mountains from 2000 to 4000 feet high, which encircled a beautiful valley teeming with vegetation. This was the celebrated valley of Soller, all but entirely occupied in its lower area by groves of Orange-trees, whilst up the sides of the surrounding mountains crept a forest, first of Olive-trees, up to an elevation of 1200 feet, then they were succeeded by a forest of Ilex and Aleppo pine, which nearly reached the summit. There were a few Lemon-trees, but only here and there. They did not appear to form an element in the cultivation.

Thus did I learn, by ocular demonstration, that the far-famed oranges of Majorca, instead of being the produce of the entire island, as generally supposed, are merely the produce of a natural orchard-house, occupying an area similar to the bottom of the crater of an extinct volcano, with sides several thousand feet high, and with only one small outlet northward to the sea. Indeed, I should have thought that the Soller valley was really an extinct volcanic crater, had it not been that the mountains, as far as I saw, were entirely limestone.

It was much warmer at Soller than at Palma or in the centre of the island, or than at Minorca, which I had previously visited, and it soon became evident to me that this valley in summer must be a natural oven, stove, hothouse. The rays of the sun being nearly perpendicular in summer penetrate just as directly into a crater valley as they would on a plain, while the heat is confined and increased by

refraction. Orange-trees are tropical trees, which rejoice
in tropical climates, so that the intense summer heat of
this shut-up valley no doubt agrees thoroughly with them,
and accounts for the excellence and precocity of the fruit.

Unfortunately the glory of this once happy valley seems
about to depart. Nearly all the oldest and finest trees, and
many of the younger ones, are dying, gradually but assuredly,
of a fell disease which I found a few years ago in the orange
forest at Mellis, in Sardinia, and to which the name of
" Secco," or dry disease, is there given. The terminal
branchlets become dry and brittle, and die. The drying
and death extends to the larger branches, and eventually to
the trunk, when the entire tree dies. I was told at Mellis
that this disease only attacked non-grafted trees—trees
grown from seed, and never or scarcely ever trees that had
been grafted. I have examined many of the Orange-trees
here, diseased and healthy, and can find no trace of their
having ever been grafted. Can this explain the death of
the trees, which threatens to ruin the country? I am
told that the value of the annual export of oranges alone
amounts to above 30,000*l.* Not knowing Spanish, and
getting on principally by signs, I was not able to discuss
these questions with the natives at Soller. A Palma gentleman, however, who owns an orange orchard at Soller, told
me that the Sardinian statement holds good in Majorca, the
diseased trees are non-grafted.

I would mention that it appears the univeral custom here
to cultivate the ground under the Orange-trees up to their
trunks, principally with vegetables. May not the ground
be thus exhausted during the summer growth of the trees,
and may not subsequent heavy dressings in winter, with a
view to remedy the exhaustion of the ground by the superficial crops, still further injure the trees?

From my experience in very many parts of the Mediterranean shores and islands I think I am warranted in saying
that the Orange-tree cannot stand wind at all, from whatever
quarter. It certainly is only successfully cultivated between

four stone walls, as at Malta, or in folds, depressions, cavities —sheltered, protected from the wind, as in Sardinia, Corfu, and Majorca. Its wonderful fertility, the immense number of fruit it produces, evidently enables a very small but sheltered locality to play a great part in the commercial production of oranges. I have often thought that good oranges might be produced commercially in England. Certainly the orange-tree is as manageable as the vine, and bears confinement equally well, only it must be submitted to great summer-heat to bear good fruit, and the expense of the necessary summer-heating might destroy profit, which, however, is great. The Soller oranges are sold at about 16s. the thousand. In England they would fetch nearly that the hundred.

The really magnificent Soller valley and mountain amphitheatre is connected by a narrow pass with a second amphitheatre, still grand and majestic, although on a smaller scale, which forms the port and harbour of Soller, about three miles from the town. The mountains which form the sides open to the extent of about a third of a mile, to let in the sea, which forms the port. The latter presents about a mile and a half in circumference. The sea entrance completes the circle. In this beautiful little port there is a small jetty, deep water, and all but perfect shelter.

The mountains slope up gradually all around to a height of about 1000 feet, and from the lighthouse which limits the western sea angle the view is indescribably beautiful; indeed, I have scarcely in any part of the Mediterranean seen anything so beautiful as the view looking landwards. The two amphitheatres of mountains, rising one above the other into the clouds, the smiling, quiet, little lake-port at our feet, with its rocky channel winding towards the sea, fill the mind of the observer with admiration. Seawards the entrance to the port widens out, and is protected by bold, frowning, precipitous headlands, which terminate abruptly in deep water. The mouth of the harbour channel is wide, and smilingly offers shelter and protection, but woe

betide the ship that went astray and struck on the frowning
rocks.   It would be heard of no more.

I spent two days on these rocks with a companion, I am
happy to say, and recognised the vegetation of the warmest
regions of the Mediterranean limestones : various species of
Euphorbia, including the dendroides, Cneorum tricoccum,
Juniper, several species; Aleppo pine, Rosemary, Lentiscus,
Smilax aspera, Alaternus, wild Olive, &c.

In both islands, near most of the houses in the country,
there are orchards of prickly Pear (*Opuntia*), evidently
cultivated for its fruit, which we think insipid ; but its
slight acidity is evidently agreeable to the inhabitants of
these regions in the latter part of summer, when apricots
and peaches are over, and grapes are not yet ripe.   I saw
but few vines cultivated for eating, only in vineyards for
wine, and that in limited areas.

In concluding this brief sketch of the spring vegetation
of Majorca, I would add that I saw but very few flowers,
cultivated or wild.   The wild ones were, no doubt, things
of the past, burnt up by the sun, for it had not rained for a
month.   I only saw a few patches of the commonest cultivated garden flowers at Palma.   In the garden of a nobleman
in the interior I found Carnations, which are much prized
and cultivated all over the Mediterranean for winter deco-
ration and for placing in the hair of the young women, Nas-
turtiums, Stocks, Delphiniums, Hollyhocks, Bengal Roses,
garden Poppies, Marigolds, white Lilies, Pansies, and that is
all.   There are a few date Palms here and there in the
island, but being useless they are not cultivated.

MINORCA.

There is a small mail steamer, of about 350 tons burden,
which leaves Barcelona every Tuesday, at 4 P.M., for the
island of Minorca.   This steamer reaches Alcudia, a port
on the north-east coast of Majorca, the next morning at five
or six, then crosses the strait which separates the two
islands, reaching Port Mahon, at the southern extremity
of Minorca, about two or three in the afternoon.   The

entire voyage from Barcelona thus occupies about twenty-
two hours. Wishing to join this steamer at Alcudia on the
Wednesday morning, I had to sleep there.

The road from Palma crosses the level plain which
occupies the centre of the island, deflecting to the north-
east. There is a railroad, recently opened—a satisfactory
evidence of incipient energy and progress, entirely constructed by Majorcan capital, which is destined to connect
the two towns. It is, however, so far, only completed to
Inca, a small agricultural town of the central region. Eight
or ten miles of road have still to be got over in the
omnibus, a species of covered cart on half-springs, holding
eight persons, including the driver, very much like an old-
fashioned English market-cart.

The country that we traversed was thickly studded with
villages and small towns, and carefully cultivated. The
rocks that occasionally showed themselves through the soil
were limestone, and the agricultural produce was principally, indeed all but entirely Cereals—Wheat and Barley;
the grain fields being dotted, at the distance of thirty or
forty feet, by Almond, Fig, Carouba, or Olive trees. It
appears that these fruit trees greatly add to the value of the
land, and, consequently, to the rent paid for it.

We found homely but clean shelter at the little *fonda* at
Alcudia, an interesting old Moorish fortified town, with ram-
parts and ditch still intact. We were roused at five the next
morning to proceed in our market-cart omnibus to the port,
about a mile distant. There we saw the Barcelona steamer
lying at ease on the green Mediterranean waters, and by 6.30
were fairly off. The Bay of Alcudia is wide, deep, sinuous,
and affords perfect shelter in deep water from all winds but
the north-east, to which it is quite exposed, as is Majorca
generally. Alcudia is only twenty-four miles distant from
Ciudadela, the capital of Minorca, which is directly opposite, at
the northern extremity of the island; whereas Port Mahon,
which occupies the south-eastern extremity, is forty-five
miles distant. The weather was fine, but there was a
heavy swell running, the remains of a former gale, so our

little steamer rolled very freely. The French steamers for Algiers pass through these straits on both journeys, and we were fortunate enough to cross the bows of one of the largest as it calmly advanced on the rolling waves like a floating castle. We were dancing about on the top of the swell in our small craft, and envied the majestic steadiness of the *Said*. This fine ship was crowded with passengers, who lined the side next to us, watching, no doubt, our erratic movements with deep commiseration.

PORT MAHON.

After rather a rough passage of six hours, at half-past two we entered the splendid harbour of Port Mahon. It is three miles in depth, and protected from every wind by its sinuous, serpentine character. Port Mahon is said, indeed, to be one of the finest harbours in the world. The town is situated on its inner and western shore on a rocky elevation, about seventy or a hundred feet above the sea level. We found very comfortable accommodation at a small *fonda*, or inn, with sash windows as in England, the cleanliness of which was beyond all praise. The entire town is clean beyond belief or description, the houses appearing to have been recently whitewashed both inside and out, and the streets being so cleansed and brushed that a pin would be seen on the ground. This extreme cleanliness continued to be the characteristic of the island wherever we went. Villages, towns, lone houses, all must be whitewashed, inside and out, every month, week, or day, for the walls are all white as snow. Indeed, I was told that whitewashing is a national tendency or craze, and that if a Minorcan gains a peseta (10*d*.) he spends a quarter of it in whitewash! I felt quite humiliated at my own individual shortcomings, and determined that when I reached home again I would imitate my Minorcan friends, and begin a vigorous and oft-renewed whitewash crusade on my own premises.

It is, no doubt, owing partly to this whitewashing mania, and to the horror of dirt which it implies, that Minorca owes its reputed healthiness and its comparative freedom from zymotic or dirt diseases. The custom gives a very peculiar character to the landscape, for as the roofs are often whitewashed as well as the walls, the houses, villages, and towns stand out in the glare of the sunshine like masses of chalk. Is this excessive cleanliness owing to Minorca having been during the greater part of the eighteenth century in the possession of the English, or is it a remains, a trace of early possession by the Moors, who reigned over it for centuries ?

Minorca is 33 miles in length and 13 in breadth where broadest; the circumference is 62 miles, and the area 300 square miles. Its longest diameter is N.W. by S.E., the latitude is 39° 47′, longitude between 3° 50′ and 4° 23′ E. It presents the character of a rocky, undulating plain, with a ridge of hills running across the island in a slanting direction, from north-west to south-east. This ridge culminates in the centre of the island near the eastern coast, at Monte Toro, about 700 feet above the sea. The southern third of the island has not even this slight barrier to the north winds. It is totally unprotected.

Vegetation.—Wishing to study the vegetation of the island I took a carriage to Ciudadela, the capital, 24 miles north of Port Mahon, making a leisurely progress one day, and returning the next. This journey proved a very pleasant and a very interesting one. The road, which passed through the centre of the island, is very good, and the covered omnibus market-cart which conveyed me and my companion was of a much better description than those we had ridden in at Majorca. It was on firm wheels, and was supplied with much better springs. These conveyances are, no doubt, the very thing for the climate of the Balearic Islands, which, during a great part of the year, is characterised by ardent sunshine, intense glare, and much dust.

The groundwork or skeleton of Minorca is a secondary limestone, but the ridge of hills that crosses the island is volcanic, and probably connected with some similar develop-

T

ment in Majorca, for the direction of the mountain chains
in the two islands is all but identical.

The weather was lovely, enchanting, as I have always
found it in the Mediterranean in the month of May, the
mid-day heat in the shade being 74° or 75°, and the night
temperature from 60° to 68°. Every plant was green,
fresh, beautiful; the sky clear, with only a few fleecy
clouds, and all Nature was bathed in glorious sunshine.

In the lower or southern third of the island there are
scarcely any trees, owing to there being no protection whatever from the sea winds from whencesoever they come. The
country, rocky and undulating, is cultivated all but entirely
with cereals. Barley quite ripe, partly cut, May 10; wheat
turning colour, and here and there a field of oats. What
was not under cereals, or fallow, was planted with broad
Beans, ripening; Potatoes, 18 inches high, in flower; and
red Clover, recently introduced, and thriving on the most
sterile limestone soils. Formerly the Vine was extensively
cultivated, and much wine made; but the oidium destroyed
the vines, and they have been replaced by cereals. The ground
is divided into small fields separated by walls three or four
feet high, and two, three, or four feet wide, made with the
stones taken out of them. There were rocky patches in
the process of formation into fields by the clearance of the
stones and rocks, and by their erection into walls.

Where the road was below the surrounding level, in slight
depressions of surface, not yet brought into cultivation, I
recognised the old familiar plants of the Mediterranean
limestone flora, the "maquis" of Corsica and Sardinia,
Lentiscus, Alaternus, Cistus, Smilax aspera, prickly Broom,
Blackberry, Honeysuckle, Asphodel, out of flower, Ferula,
the same, Convolvulus, variegated Thistle, Geranium, Cine-
raria maritima, the universal Shepherd's purse, Gladiolus,
Chrysanthemum segetum, Myrtle, Rosemary, Thyme, Rue,
&c. Evidently this part of Minorca was, once upon a
time, covered with this kind of vegetation, and if left to
itself for a few years would soon return to the wild state,

and be covered by these plants, the denizens of the warmer regions of the Mediterranean, on limestone soils.

About nine miles from Port Mahon we reached the base of the intersecting ridge of hills, at the foot of which, lightly protected from the north-east winds, vegetation became more luxuriant, Fig-trees and small Olive-trees appearing. Up to that time all watercourses had been quite dry; they were clearly mere winter torrents, full for an hour or a day, and then dry. Here we saw a small watercourse, a mere rivulet, all but dry, fringed with Tamarisks. I looked for the Oleander, the usual companion of the Tamarisk along the beds of streams and torrents in Algeria (only 190 miles distant), but I did not see any. The Ilex or evergreen Oak, a Mediterranean tree in schistic, granitic soils, also appeared.

The road entered a depression in the hills, and the nature of the soil and rock changed, becoming schistic, volcanic. The Ilex were more numerous and larger, and the brushwood was principally composed of Arbutus, Mediterranean Heath, Caluna vulgaris, and Cistus or rock Rose, as is usually the case with such soils in the Mediterranean. After winding for a few miles through these low hills covered with a thicket of Ilex and brushwood, we again emerged into a calcareous rocky plain, which reproduced the vegetation and cultivation of the southern region already described, reaching Ciudadela for dinner.

Ciudadela.—Ciudadela is a very clean " whitewashed" little town, with a population of 8000. It is the legal and Government capital of the island, Port Mahon being the commercial and military one, with a population of 18,000. The fortifications still extant are of Moorish origin. Here we got very comfortable and exquisitely clean accommodation, at an inn kept by a man who had been many years cook on board a United States man-of-war, and who received me and my companion most cordially. There was a public garden at Ciudadela, a *rambla,* as it is called in the north of Spain, and I made a note of what I found—viz., white Lily,

T 2

red Valerian, Nasturtium, Wallflower, Holyhock, centifolia Rose, Bengal Rose, Cytisus, Poppy, Antirrhinum, Adonis, fancy Pelargoniums, very poor; Oak-leaved geranium, sweet Pea, Phlox, Carnation, Delphinium, and two or three Palms, thirty feet high, the only ones I saw in Minorca. All these were in full flower, and were thought by the natives to constitute a marvellously beautiful and choice garden. They are found in all gardens of the Mediterranean islands and shores, constituting the principal spring garden flora. I presume they occupy the same position of honour in " cottage gardens" all over the temperate regions of the world.

Where the Oranges Grow.—Both at Port Mahon and at Ciudadela there were plenty of oranges to be had, and yet after crossing all but the entire island I had not seen a single orange tree or shrub. Where could they come from was the question I asked. I was told that they all came from one garden or orchard, an hour's ride from the village of Ferarias, a few miles from Ciudadela, near the main road ; so we determined to stop on our way back and examine this wonderful orchard, the existence of which on an exposed island like Minorca puzzled me greatly.

Two hours' drive brought us next day to Ferarias ; we mounted donkeys, and scrambled over rocky hilly paths for an hour in the glare of the sun, without seeing the vestige of a garden, of trees, or of anything else but the field and occasional hedge or wall vegetation already described. We were nearing the sea also, and between us and it there seemed to be nothing but stony eminences, declivities, and stone-enclosed fields planted with cereals. All at once our guide pointed to a cleft in the rocks on our left, the beginning of a sinuous depression which, at 100 yards distance, seemed like the depressed bed of a river, occupying a gorge with cliffs on both sides. Our donkeys entered the cleft between two rocks, twenty feet apart, by a narrow path, and we at once found ourselves in cool welcome shade, in the midst of most luxuriant vegetation. The path and the valley or gorge soon expanded, and we discovered that

we were truly in a "happy valley." We had been transported in two minutes from burnt-up, scorched rocks, with a
scanty spring vegetation, to a tropical forest of Pomegranate-
trees, quite timber trees, Orange, Lemon, Fig, and Olive-trees,
growing in the greatest possible luxuriance. The ground
and rocks were covered with rank grasses, with Buttercups,
Periwinkle, Geranium, Sage, Docks, Ivy, Clematis, and with
Capillus veneris, Scolopendrium, and Polypodium vulgare
ferns.

This wonderful valley is a mere cleft, rent, depression in
the calcareous rocks of this part of the island.   It is about
1½ mile long, sinuous, of variable width, from 50 to 300
feet, with cliffs about 150 feet high on the west side, and
about 70 feet or 100 feet on the east.   It reminded me
very much of the sinuous cleft, depression, gorge, or valley
which constitutes the harbour of Port Mahon.   But the
Ferarias valley has no communication with the sea, is much
narrower, and its cliffs are higher and more precipitous; its
serpentine direction helps to protect it from wind, whencesoever it blows, as is likewise the case with the harbour of
Port Mahon.

The central, widest, and most sheltered part of the valley
is entirely occupied by large healthy Orange-trees, with
trunks two, three, and four feet in circumference, grown as
timber trees, and they have not suffered as yet from the *secco*,
although I could not discover that they had been grafted.
The farmer told us that this orchard not only supplied all
Minorca with oranges, but enabled the owners to export
large quantities every year.   They had just sent off 50,000,
and yet the entire area occupied by the orange-trees did
not amount to more than a few acres.

I learnt here, as in Majorca, that the size of the fruit
depends, not on the species of the tree, but on the number
of fruit the tree is allowed to bear.   If few they are large,
if numerous small, as is the case with fruit in general.   The
contrary opinion reigns all over the western Mediterranean.
The large Majorca oranges, which are always sold at 20c.

apiece, twopence, are supposed to be a distinct species. As
usual with all fruit, not only is the size thus increased, but
the flavour seems to improve along with the size.

Thus in Minorca the Ferrarias orange orchard reproduces
in a different way the history of the Soller Valley orange
orchard in Majorca. The one is a gorge, or fault in the
rocks, thoroughly protected from winds, as the other is a
kind of crater-like amphitheatre, also inaccessible to wind.
Both are natural stoves, orchard-houses, offering complete
protection from every wind, which seems an indispensable
condition to orange culture.

Importance of Protection.—The luxuriant vegetation of
these sheltered localities reproduces what I have found all over
the Mediterranean. Exposure to north winds is quite incom-
patible with subtropical vegetation anywhere in the Mediterranean, even on its southern shore. The north winds sweep right
down into the desert, whenever they meet with no barrier on
their way, and peel the rocks, leaving nothing but pines and
aromatic shrubs. Such, however, is not the case with south
winds, if there is protection from the north, as evidenced by
the Genoese Riviera from Nice to Genoa. In the more
sheltered regions lemon-trees clothe the hills down to the
sea-shore, and even orange-trees can grow and flourish near
the shore if they are protected by a high wall. Thus once
more Majorca and Minorca illustrate the lesson of the inestimable value of protection from north winds in estimating
climate, and of the extreme influence that it exercises on
vegetation in subtropical as well as in northern regions.
We see repeated in these islands the facts brought to light
in every other region of the Mediterranean shores and
islands that I have studied and described elsewhere.

The climate of the Balearic Islands is an insular one,
the hygrometer constantly showing the atmosphere to contain much moisture, and often to be saturated therewith.
This is a necessary result of the winds that reach the island
having to cross a considerable expanse of sea whencesoever
they come. The average amount of rain is about fifteen
inches, falling principally at the autumnal and spring

equinoxes. This rain is generally precipitated by the contact of the moisture-laden winds with the mountain chain that runs from west to east along the northern shore of the island. The rains are very variable; some years very little falls, the torrents, springs, and wells become dry, and great suffering ensues to man and beast, as in 1847 and 1849. In other years torrents of rain fall in the mountains, pour down into the plains and cause great devastation. Such was the case in Majorca in 1850, the year after the drought; several hundred houses were destroyed. The wide torrent beds filled with rounded rocks and boulders both on the north and south sides of the mountains testify to these heavy rains. In winter there are frequently violent winds from all quarters, but principally from the north-east and north-west. The cretaceous mountain range which runs in Majorca along the north shore, although sheltering the island from the north, as stated, affords little protection from these winds.

There are no rivers in either Majorca or Minorca, merely mountain torrents and rivulets filled temporarily by the rains. In the years in which the usual amount of rain falls there are springs and wells in most parts of the islands, but they dry up in dry seasons. As in Malta, however, rain is everywhere stored, in large subterranean tanks supplied by the water-shed of the roofs. The Moors, who long occupied the Balearic Islands, in the Middle Ages, have left at Palma and elsewhere great works of this description.

The climate is mild and temperate in winter, the thermometer seldom falling below the freezing point, warm in summer, but less so than that of the neighbouring Spanish mainland. The annual mean at Palma is 65° F. against 66°·6 for Valentia. The coast of Algeria, 150 miles distant, has the same annual temperature, about 65°. The lowest temperature at Palma is in February, and the mean of the month is 45° F., the highest is in August, and the mean of the month is 92°. Valentia has a winter minimum of 35°·5 for January, and a summer maximum of 97°, the one lower, the other higher than those of Palma.

The pathology of the Balearic Islands is very similar to that of the islands and large towns of the Mediterranean, both in the past and in the present. Palma, the capital of Majorca (50,000 inhabitants), is, and has been for centuries, a populous town. In the Middle Ages it used to be ravaged by the plague three, four, or five times each century. In 1653, 406 houses remained uninhabited in Palma, the occupants and owners having all died of the plague; 2198 men died out of 3015 attacked; 3293 women out of 5797; 3981 children out of 4634! Since then plague has disappeared, but there have been all but equally serious epidemics of malignant fevers, malignant sore throats (diphtheria?), yellow fever (1821), and cholera.

I mention these facts to show that the health history of Palma has been the same in the past, as that of Malaga, Barcelona, Marseilles. However healthy the position of a southern city may be. even when like Palma, built all but in the middle of the sea, on a declivity, open to all the winds that blow, free from any external source of contamination, if the laws of town hygiene are neglected, if the streets are narrow and dirty, the drainage left to fester, all the pestilences to which the human race is subject run riot and reign supreme, they keep destroying life constantly, but imperceptibly, and every now and then they sweep off the population by wholesale.

Palma and also Port Mahon, the capital of Minorca, are now better prepared, however, to meet pestilence than formerly. I found them both exquisitely clean, whitewashed, and brushed, until scarcely a speck of dirt could be seen or found in either. I was told that Palma had been transformed within a few years by a vigorous "tyrannical" mayor. Having first secured the assistance of his town council he had made a vigorous and successful crusade against all the dirt habits of the south, pulling down houses, forbidding people to sit in the streets, in the southern way, forbidding mendicity, and in fifty other ways improving the looks and sanitary state of the city. There is one point respecting which he has not succeeded in modifying the convictions and habits of his fellow-citizens, if I may judge by the state of the hotel where I resided, and by that of several houses that I casually visited.

In Southern Europe the entire population seem to be ignorant of the fact that the fermentation of human fæcal accumulations, especially in hot weather, generates pestilential gases, which give rise to putrid or diphtherieal sore throats, typhoid fever, and other serious diseases. Then the closets are constantly built in the centre of the inhabited houses, without traps or precautions of any kind, and the houses themselves are constantly permeated with offensive odours, indicating the danger. So it was in the Middle Ages everywhere, and this, no doubt, was one of the principal causes of town unhealthiness and mortality.

In our times the diseases that reign in the Balearic Islands are those of the other similar regions of the Mediterranean—intermittent fever, rheumatism, catarrhal fever, typhoid, diarrhœa, dysentery, &c. Pulmonary Consumption is not uncommon in Palma. (Fernando Weyler, "Topographie Physico-Médicale des Iles Balearic." Palma, 1854.) It is the same in all the large close towns of the Mediterranean. Intermittent fever is very common in summer and in autumn. There are some salt water marshes on the S.E. coast near Alcudia, but with that exception the islands are quite devoid of marshes of any description. Yet intermittent fever develops itself all over the island irrespective of any possible malarious influence. This is another illustration added to those I have given in my work on "Climatology" gathered from various parts of the Mediterranean, of the fact that mere chill in a tropical or subtropical climate will produce intermittent fever in constitutions predisposed by great previous heat, without any malarious influence. This we are now seeing exemplified on a large scale by our troops in Cyprus, an island no doubt very similar to Majorca (1878). Troops disembarked on any island in the Mediterranean in mid-summer, and treated as they have been, as if they had merely been at Aldershot, would have been equally sickly.

The conclusions at which I arrived, after this searching investigation of the two principal islands of the Balearic Islands, (I had not leisure to visit Ivica,) are, that they cannot be recommended to invalids as winter residences. The climate is mild, it is true, in winter, nearly as mild apparently as that of Algiers, but the air is too moist, the winds are too constant and too vehement, and the storms and rain too violent in rainy seasons. The mountain range is not sufficiently wide at its base to exhaust and concentrate these storms and to protect the island from their influence. Indeed, there is no one condition that makes these islands preferable to the mainland—Malaga, Murcia, Elche, Alicante, Valentia (see my work on the Mediterranean, Chapter IX., Spain). I must add, there is no accommodation for invalids. The inns at Palma are simply vile, and there is no other refuge for them in this large city. Invalids seem to be neither wanted nor cared for. The Spanish invalids must be taken in by friends and relations, for I failed to discover any place to which they could go. Out of Palma in the numerous villages there is simply no accommodation whatever. Port Mahon does contain a decent posada or inn, and has been so long in the occupation of the English that, like Corfu, it retains many of the evidences of our civilisation. But there is nothing there to attract the stranger except the mildness of the climate, and the singular long sinuous port, formed by a fault or dislocation between the calcareous formation (lias) which constitutes the backbone of the island, and some tertiary strata. There is, however, a fine cathedral and a grand organ, presented by an invalid Englishman. On this organ I heard Italian Opera music played, as church and anthem music, magnificently. The cathedral at Palma is also a very grand and imposing Gothic edifice, a monument of the Middle Ages, not yet finished.

The Balearic Islands seem to me merely suited for a spring tour in April and May, such as I made. Then they are truly delightful, and well repay all the trouble of the journey.

INDEX.

THE END.

PRINTED BY BALLANTYNE AND HANSON
LONDON AND EDINBURGH

DR. JAMES HENRY BENNET'S WORKS.

I.

Third Library Edition, Revised, 8vo, pp. 260, 7s.
Cheap Reprint, 12mo, pp. 276, 2s. 6d.

NUTRITION IN HEALTH AND DISEASE.

A CONTRIBUTION TO HYGIENE AND TO CLINICAL MEDICINE.

II.

*1 vol. 8vo, pp. 654. With Eight Chromo-lithographed Maps, Frontispiece, and
Forty Wood Engravings. Fifth Edition. Price 12s. 6d.*

WINTER AND SPRING ON THE SHORES AND ISLANDS OF THE MEDITERRANEAN;

OR,

*THE GENOESE RIVIERAS, ITALY, SPAIN, CORFU, GREECE, THE ARCHI-
PELAGO, CONSTANTINOPLE, CORSICA, SICILY, SARDINIA, MALTA,
ALGERIA, TUNIS, SMYRNA, ASIA MINOR, AND BIARRITZ AS WINTER
CLIMATES.*

" Dr. Henry Bennet was one of the earliest to advocate the adoption of hygienic
measures in the health resorts of southern Europe, and he has never ceased to
insist on the importance of such measures, especially for invalids."—*Medical Times
and Gazette.*

" The interesting nature of its contents, the mingling of science with general
information, render it equally suitable for the general reader or the physician."—
Edinburgh Medical Journal.

" A most instructive and delightful volume, creditable alike to the head and the
heart of the writer."—*Lancet.*

" A most pleasantly written and instructive account."—*British Medical Journal.*

" The perusal of Dr. Henry Bennet's book can be conscientiously recommended,
therefore, not only for its delightful descriptions of nature and of natural phenomena in some of their grandest and most gorgeous phases; not only for its pleasant
narratives of local incidents and usages, and social peculiarities, its instructive
and clearly-written scientific notes, its botanical and horticultural elucidations,
and many other matters to which reference has been made; but on the broader and
higher ground of the practical good which it promises to bring to the sufferer.
But whatever the object with which it may be taken up, it will certainly not be
laid down before every page of it has been read, or even re-read. It exhibits the
Mediterranean graphically and truthfully from the Pillars of Hercules to the
shattered temples of Greece and her colonies; and as setting forth the impressions
which those charming scenes made on a cultivated mind, amply vindicates the
favour with which it has been received."—*Morning Post*, Nov. 30, 1874.

III.

Third Edition, 8vo, pp. 286, 7s. 6d.

THE TREATMENT OF PULMONARY CONSUMPTION BY HYGIENE, CLIMATE AND MEDICINE,

WITH AN APPENDIX ON THE SANITARIA OF THE UNITED STATES, SWITZERLAND AND THE BALEARIC ISLES.

" In concluding our observations we feel no hesitation in saying that both the
manner in which the book is written, and the matter which the book contains,
render it a desirable volume to place on the shelves of any medical man's library.
. . . . Any physician may take it up with every feeling of confidence that the
views enunciated by the author will be found to be able, honest, and orthodox."—
British and Foreign Medico-Chirurgical Review, March, 1874.

IV.

Fourth Edition, 8vo, pp. 600.   Price 12s.

A PRACTICAL TREATISE ON UTERINE DISEASES.

*This Work has gone through four English editions and five American editions ;
it has been twice translated into French, and once into German.*

SECOND EDITION, 1848.

" This complete and perfect specimen of medical investigation, which combines those rare
concomitants in the literature of our profession—an almost new subject treated in a scientific
yet agreeable manner."—*Provincial Journal.*

" The fact of a second edition having been called for is sufficient proof of the estimation in
which the author is held, and confirms us in the favourable opinion we expressed of the former
edition of this treatise. A vast deal of important matter has been added, so that, although
nominally a second edition, it is in reality a new work. It only remains for us to express our
deep admiration for the talents of the author, and our sense of obligation for the great benefit
he has conferred on this department of practical medicine."—*Dublin Quarter'y Journal.*

" We presume there are few medical men, whose practice embraces the diseases in question,
who will not possess themselves of the volume, which we have no hesitation in pronouncing to
be the most original, as well as the most complete upon the subject, that it has fallen to our
lot to peruse."—*Dublin Medical Press* (1849).

" We are firmly of opinion that in proportion as a knowledge of uterine diseases becomes
more appreciated, this work will be proportionately established as a text-book in the profession." - *Lancet* (1850).

FOURTH EDITION, 1862.

" A most valuable contribution to medical science. It may be fairly said that Dr. Henry
Bennet's writings have given a more accurately determined impulse to the study and understanding of the diseases to which females are liable than those of any other author of the
present century."—*Lancet.*

V.

Preparing for Publication. Second Edition, 1 vol. 8vo.

A REVIEW OF THE PRESENT STATE OF UTERINE PATHOLOGY.

" This essay is remarkable, not alone for excellent composition, admirable arrangement, and
lucid reasoning, but likewise for the calm, candid, and gentlemanlike manner in which the
author has dealt with his opponents."—*Dublin Quarterly Journal.*

London: J. & A. CHURCHILL, New Burlington Street.

DR. JAMES HENRY BENNET'S FRENCH WORKS.

I. RECHERCHES sur le TRAITEMENT de la PHTHISIE
PULMONAIRE par l'Hygiène, les Climats, et la Médecine, dans ses
rapports avec les doctrines modernes. 1 vol. 8vo, pp. 221. 5s. 1874.

II. LA CORSE et la SARDAIGNE; ÉTUDE de VOYAGE et de
CLIMATOLOGIE. 1 vol. 12mo, pp. 254. Trois cartes chromo-litho-
graphiées. 3s. 6d.
Paris : R. ASSELIN, Place de l'Ecole de Médecine.

1 vol. 8vo, pp. 592.

III. TRAITÉ PRATIQUE DE L'INFLAMMATION DE
L'UTERUS, DE SON COL ET DE SES ANNEXES; et des rapports
de cette Inflammation avec les autres affections utérines. Première
édition anglaise, 1845, Londres. Quatrième, 1862. 1 vol. in-8 de
600 pp. Première traduction française, par le docteur ARAN, Paris, 1852.
Seconde traduction française, par le docteur MICHEL PETER, Professeur
à la Faculté de Médecine de Paris. Paris, 1864 : R. Asselin.

London, *New Burlington Street.*
October, 1878.

SELECTION

FROM

MESSRS J. & A. CHURCHILL'S

General Catalogue

COMPRISING

ALL RECENT WORKS PUBLISHED BY THEM

ON THE

ART AND SCIENCE

OF

MEDICINE

INDEX

THE PRACTICE OF SURGERY :
a Manual by THOMAS BRYANT, F.R.C.S., Surgeon to Guy's Hospital.
Third Edition, 2 vols., crown 8vo, with 672 Engravings. [1878]

THE PRINCIPLES AND PRACTICE OF SURGERY,
by WILLIAM PIRRIE, F.R.S.E., Professor of Surgery in the University
of Aberdeen. Third Edition, 8vo, with 490 Engravings, 28s. [1873]

A SYSTEM OF PRACTICAL SURGERY,
by Sir WILLIAM FERGUSSON, Bart., F.R.C.S., F.R.S. Fifth Edition,
8vo, with 463 Engravings, 21s. [1870]

OPERATIVE SURGERY,
by C. F. MAUNDER, F.R.C.S., Surgeon to the London Hospital.
Second Edition, post 8vo, with 164 Engravings, 6s. [1872]

BY THE SAME AUTHOR.

SURGERY OF THE ARTERIES :
Lettsomian Lectures for 1875, on Aneurisms, Wounds, Hæmorrhages,
&c. Post 8vo, with 18 Engravings, 5s. [1875]

THE SURGEON'S VADE-MECUM,
a Manual of Modern Surgery, by ROBERT DRUITT. Eleventh Edition,
fcap. 8vo, with 369 Engravings, 14s. [1878]

THE SCIENCE AND PRACTICE OF SURGERY :
a complete System and Textbook by F. J. GANT, F.R.C.S., Senior Surgeon to the Royal Free Hospital. 8vo, with 470 Engravings, 24s. [1871]

OUTLINES OF SURGERY AND SURGICAL PATHOLOGY,
including the Diagnosis and Treatment of Obscure and Urgent
Cases, and the Surgical Anatomy of some Important Structures and
Regions, by F. LE GROS CLARK, F.R.S., Consulting Surgeon to St.
Thomas's Hospital. Second Edition, Revised and Expanded by the
Author, assisted by W. W. WAGSTAFFE, F.R.C.S., Assistant-Surgeon
to St. Thomas's Hospital. 8vo, 10s. 6d. [1872]

CLINICAL AND PATHOLOGICAL OBSERVATIONS IN INDIA,
by Sir J. FAYRER, K.C.S.I., M.D., F.R.C.P. Lond., F.R.S.E., Honorary
Physician to the Queen. 8vo, with Engravings, 20s. [1873]

TREATMENT OF WOUNDS :
Clinical Lectures, by SAMPSON GAMGEE, F.R.S.E., Surgeon to the
Queen's Hospital, Birmingham. Crown 8vo, with Engravings, 5s. [1878]

BY THE SAME AUTHOR,

FRACTURES OF THE LIMBS
and their Treatment. 8vo, with Plates, 10s. 6d. [1871]

THE FEMALE PELVIC ORGANS,
their Surgery, Surgical Pathology, and Surgical Anatomy, in a
Series of Coloured Plates taken from Nature: with Commentaries,
Notes, and Cases, by HENRY SAVAGE, M.D. Lond., F.R.C.S., Consulting
Officer of the Samaritan Free Hospital. Third Edition, 4to, £1 15s.
[1875]

SURGICAL EMERGENCIES
together with the Emergencies attendant on Parturition and the Treatment of Poisoning : a Manual for the use of General Practitioners, by WILLIAM P. SWAIN, F.R.C.S., Surgeon to the Royal Albert Hospital, Devonport. Second Edition, post 8vo, with 104 Engravings, 6s. 6d. [1876]

TRANSFUSION OF HUMAN BLOOD:
with Table of 50 cases, by Dr. ROUSSEL, of Geneva. Translated by CLAUDE GUINNESS, B.A. With a Preface by SIR JAMES PAGET, Bart. Crown 8vo, 2s. 6d. [1877]

ILLUSTRATIONS OF CLINICAL SURGERY,
consisting of Coloured Plates, Photographs, Woodcuts, Diagrams, &c., illustrating Surgical Diseases, Symptoms and Accidents ; also Operations and other methods of Treatment. By JONATHAN HUTCHINSON, F.R.C.S., Senior Surgeon to the London Hospital. In Quarterly Fasciculi, 6s. 6d. each. Fasciculi I to X bound, with Appendix and Index, £3 10s. [1876-8]

PRINCIPLES OF SURGICAL DIAGNOSIS
especially in Relation to Shock and Visceral Lesions, by F. LE GROS CLARK, F.R.C.S., Consulting Surgeon to St. Thomas's Hospital. 8vo, 10s. 6d. [1870]

MINOR SURGERY AND BANDAGING:
a Manual for the Use of House-Surgeons, Dressers, and Junior Practitioners, by CHRISTOPHER HEATH, F.R.C.S., Surgeon to University College Hospital, and Holme Professor of Surgery in University College. Fifth Edition, fcap 8vo, with 86 Engravings, 5s. 6d. [1875]

BY THE SAME AUTHOR,
INJURIES AND DISEASES OF THE JAWS:
JACKSONIAN PRIZE ESSAY. Second Edition, 8vo, with 164 Engravings, 12s. [1872]

BY THE SAME AUTHOR.
A COURSE OF OPERATIVE SURGERY:
with 20 Plates drawn from Nature by M. LÉVEILLÉ, and coloured by hand under his direction. Large 8vo. 40s. [1877]

HARE-LIP AND CLEFT PALATE,
by FRANCIS MASON, F.R.C.S., Surgeon and Lecturer on Anatomy at St. Thomas's Hospital. With 66 Engravings, 8vo, 6s. [1877]

BY THE SAME AUTHOR,
THE SURGERY OF THE FACE:
with 100 Engravings. 8vo, 7s. 6d. [1878]

FRACTURES OF THE LOWER END OF THE RADIUS,
Fractures of the Clavicle, and on the Reduction of the Recent Inward
Dislocations of the Shoulder Joint. By ALEXANDER GORDON, M.D.,
Professor of Surgery in Queen's College, Belfast. With Engravings,
8vo, 5s. [1875]

DISEASES AND INJURIES OF THE EAR,
by W. B. DALBY, F.R.C.S., M.B., Aural Surgeon and Lecturer on
Aural Surgery at St. George's Hospital. Crown 8vo, with 21 Engravings, 6s. 6d. [1873]

AURAL SURGERY ;
A Practical Treatise, by H. MACNAUGHTON JONES, M.D., Surgeon to
the Cork Ophthalmic and Aural Hospital. With 46 Engravings,
crown 8vo, 5s. [1878.]

BY THE SAME AUTHOR,
ATLAS OF DISEASES OF THE MEMBRANA TYMPANI.
In Coloured Plates, containing 62 Figures, with Text, crown 4to, 21s.
[1878]

THE EAR:
its Anatomy, Physiology, and Diseases. A Practical Treatise, by
CHARLES H. BURNETT, A.M., M.D., Aural Surgeon to the Presby-
terian Hospital, and Surgeon in Charge of the Infirmary for Diseases
of the Ear, Philadelphia. With 87 Engravings, 8vo, 18s. [1877]

EAR AND THROAT DISEASES.
Essays by LLEWELLYN THOMAS, M.D., Surgeon to the Central
London Throat and Ear Hospital. Post 8vo, 2s. 6d. [1878]

CLUBFOOT :
its Causes, Pathology, and Treatment: Jacksonian Prize Essay by WM.
ADAMS, F.R.C.S., Surgeon to the Great Northern Hospital. Second
Edition, 8vo, with 106 Engravings and 6 Lithographic Plates, 15s. [1873]

ORTHOPÆDIC SURGERY :
Lectures delivered at St. George's Hospital, by BERNARD E. BROD-
HURST, F.R.C.S., Surgeon to the Royal Orthopædic Hospital. Second
Edition,8vo, with Engravings, 12s. 6d. [1876]

OPERATIVE SURGERY OF THE FOOT AND ANKLE,
by HENRY HANCOCK, F.R.C.S., Consulting Surgeon to Charing Cross
Hospital. 8vo, with Engravings, 15s. [1873]

THE TREATMENT OF SURGICAL INFLAMMATIONS
by a New Method, which greatly shortens their Duration, by FURNEAUX
JORDAN, F.R.C.S., Professor of Surgery in Queen's College, Birmingham. 8vo, with Plates, 7s. 6d. [1870]

BY THE SAME AUTHOR,
SURGICAL INQUIRIES.
With numerous Lithographic Plates. 8vo, 5s. [1873]

ORTHOPRAXY :
the Mechanical Treatment of Deformities, Debilities, and Deficiencies of
the Human Frame, by H. HEATHER BIGG, Associate of the Institute of
Civil Engineers. Third Edition, with 319 Engravings, 8vo, 15s. [1877]

ORTHOPÆDIC SURGERY:
and Diseases of the Joints. Lectures by LEWIS A. SAYRE, M.D., Professor of Orthopædic Surgery, Fractures and Dislocations, and Clinical Surgery, in Bellevue Hospital Medical College, New York. With 274 Wood Engravings, 8vo, 20s. [1876]

INTERNAL ANEURISM:
Its Successful Treatment by Consolidation of the Contents of the Sac. By T. JOLIFFE TUFNELL. F.R.C.S.I., President of the Royal College of Surgeons in Ireland. With Coloured Plates. Second Edition, royal 8vo, 5s. [1875]

DISEASES OF THE RECTUM,
by THOMAS B. CURLING, F.R.S., Consulting Surgeon to the London Hospital. Fourth Edition, Revised, 8vo, 7s. 6d. [1876]

BY THE SAME AUTHOR,
DISEASES OF THE TESTIS, SPERMATIC CORD, AND SCROTUM.
Third Edition, with Engravings, 8vo, 16s. [1878]

STRICTURE OF THE URETHRA
and Urinary Fistulæ; their Pathology and Treatment: Jacksonian Prize Essay by Sir HENRY THOMPSON, F.R.C.S., Emeritus Professor of Surgery to University College. Third Edition, 8vo, with Plates, 10s. [1869]

BY THE SAME AUTHOR,
PRACTICAL LITHOTOMY AND LITHOTRITY;
or, An Inquiry into the best Modes of removing Stone from the Bladder. Second Edition, 8vo, with numerous Engravings. 10s. [1871]

ALSO,
DISEASES OF THE URINARY ORGANS:
(Clinical Lectures). Fourth Edition, 8vo, with 2 Plates and 59 Engravings, 12s. [1876]

ALSO,
DISEASES OF THE PROSTATE:
their Pathology and Treatment. Fourth Edition, 8vo, with numerous Plates, 10s. [1873]

ALSO,
THE PREVENTIVE TREATMENT OF CALCULOUS DISEASE
and the Use of Solvent Remedies. Second Edition, fcap. 8vo, 2s. 6d. [1876]

STRICTURE OF THE URETHRA,
and other Diseases of the Urinary Organs, by REGINALD HARRISON, F.R.C.S., Surgeon to the Liverpool Royal Infirmary. With 10 plates. 8vo, 7s. 6d. [1878.]

LITHOTOMY AND EXTRACTION OF STONE
from the Bladder, Urethra, and Prostate of the Male, and from the Bladder of the Female, by W. POULETT HARRIS, M.D., Surgeon-Major H.M. Bengal Medical Service. With Engravings, 8vo, 10s. 6d. [1876]

THE SURGERY OF THE RECTUM:

Lettsomian Lectures by HENRY SMITH, F.R.C.S., Professor of Surgery in King's College, Surgeon to King's College Hospital. Fourth Edition, fcap. 8vo, 5s.   [1876]

FISTULA, HÆMORRHOIDS, PAINFUL ULCER,

Stricture, Prolapsus, and other Diseases of the Rectum: their Diagnosis and Treatment, by WM. ALLINGHAM, F.R.C.S., Surgeon to St. Mark's Hospital for Fistula, &c. Second Edition, 8vo, 7s.   [1872]

KIDNEY DISEASES, URINARY DEPOSITS,

and Calculous Disorders by LIONEL S. BEALE, M.B., F.R.S., F.R.C.P., Physician to King's College Hospital. Third Edition, 8vo, with 70 Plates, 25s.   [1868]

DISEASES OF THE BLADDER,

Prostate Gland and Urethra, including a practical view of Urinary Diseases, Deposits and Calculi, by F. J. GANT, F.R.C.S., Senior Surgeon to the Royal Free Hospital. Fourth Edition, crown 8vo, with Engravings, 10s. 6d.   [1876]

RENAL DISEASES:

a Clinical Guide to their Diagnosis and Treatment by W. R. BASHAM, M.D., F.R.C.P., late Senior Physician to the Westminster Hospital. Post 8vo, 7s.   [1870]

BY THE SAME AUTHOR,

THE DIAGNOSIS OF DISEASES OF THE KIDNEYS,

with Aids thereto. 8vo, with 10 Plates, 5s.   [1872]

THE REPRODUCTIVE ORGANS

in Childhood, Youth, Adult Age, and Advanced Life (Functions and Disorders of), considered in their Physiological, Social, and Moral Relations, by WILLIAM ACTON, M.R.C.S. Sixth Edition, 8vo, 12s.
                                                                   [1875]

URINARY AND REPRODUCTIVE ORGANS:

their Functional Diseases, by D. CAMPBELL BLACK, M.D., L.R.C.S. Edin. Second Edition. 8vo, 10s. 6d.   [1875]

LECTURES ON SYPHILIS,

and on some forms of Local Disease, affecting principally the Organs of Generation, by HENRY LEE, F.R.C.S., Surgeon to St. George's Hospital. With Engravings, 8vo, 10s.   [1875]

SYPHILITIC NERVOUS AFFECTIONS:

Their Clinical Aspects, by THOMAS BUZZARD, M.D., F.R.C.P. Lond., Physician to the National Hospital for Paralysis and Epilepsy. Post 8vo, 5s.   [1874]

SYPHILIS:

Harveian Lectures, by J. R. LANE, F.R.C.S., Surgeon to, and Lecturer on Surgery at, St. Mary's Hospital; Consulting Surgeon to the Lock Hospital. Fcap. 8vo, 3s. 6d.   [1878]

PATHOLOGY OF THE URINE,
including a Complete Guide to its Analysis, by J. L. W. THUDICHUM, M.D., F.R.C.P. Second Edition, rewritten and enlarged, with Engravings, 8vo, 15s. [1877]

GENITO-URINARY ORGANS, INCLUDING SYPHILIS:
A Practical Treatise on their Surgical Diseases, designed as a Manual for Students and Practitioners, by W. H. VAN BUREN, M.D., Professor of the Principles of Surgery in Bellevue Hospital Medical College, New York, and E. L. KEYES, M.D., Professor of Dermatology in Bellevue Hospital Medical College, New York. Royal 8vo, with 140 Engravings, 21s. [1874]

HISTOLOGY AND HISTO-CHEMISTRY OF MAN:
A Treatise on the Elements of Composition and Structure of the Human Body, by HEINRICH FREY, Professor of Medicine in Zurich. Translated from the Fourth German Edition by ARTHUR E. J. BARKER, Assistant-Surgeon to University College Hospital. And Revised by the Author. 8vo, with 608 Engravings, 21s. [1874]

HUMAN PHYSIOLOGY:
A Treatise designed for the Use of Students and Practitioners of Medicine, by JOHN C. DALTON, M.D., Professor of Physiology and Hygiene in the College of Physicians and Surgeons, New York. Sixth Edition, royal 8vo, with 316 Engravings, 20s. [1875]

HANDBOOK FOR THE PHYSIOLOGICAL LABORATORY,
by E. KLEIN, M.D., F.R.S., Assistant Professor in the Pathological Laboratory of the Brown Institution, London; J. BURDON-SANDERSON, M.D., F.R.S., Professor of Practical Physiology in University College, London; MICHAEL FOSTER, M.D., F.R.S., Prælector of Physiology in Trinity College, Cambridge; and T. LAUDER BRUNTON, M.D., F.R.S., Lecturer on Materia Medica at St. Bartholomew's Hospital; edited by J. BURDON-SANDERSON. 8vo, with 123 Plates, 24s. [1873]

PRACTICAL HISTOLOGY:
By WILLIAM RUTHERFORD, M.D., Professor of the Institutes of Medicine in the University of Edinburgh. Second Edition, with 63 Engravings. Crown 8vo (with additional leaves for notes), 6s. [1876]

THE MARRIAGE OF NEAR KIN,
Considered with respect to the Laws of Nations, Results of Experience, and the Teachings of Biology, by ALFRED H. HUTH. 8vo, 14s. [1877]

MANUAL OF ANTHROPOMETRY:
A Guide to the Measurement of the Human Body, containing an Anthropometrical Chart and Register, a Systematic Table of Measurements, &c. By CHARLES ROBERTS, F.R.C.S., late Assistant Surgeon to the Victoria Hospital for Children. With numerous Illustrations and Tables. 8vo, 6s. 6d. [1878]

§

PRINCIPLES OF HUMAN PHYSIOLOGY,
by W. B. CARPENTER, C.B., M.D., F.R.S. Eighth Edition by HENRY
POWER, M.B., F.R.C.S., Examiner in Natural Science, University of
Oxford, and in Natural Science and Medicine, University of Cambridge.
8vo, with 3 Steel Plates and 371 Engravings, 31s. 6d.   [1876]

STUDENTS' GUIDE TO HUMAN OSTEOLOGY,
By WILLIAM WARWICK WAGSTAFFE, F.R.C.S., Assistant-Surgeon
and Lecturer on Anatomy, St. Thomas's Hospital. With 23 ¿Plates
and 66 Engravings. Fcap. 8vo, 10s. 6d.   [1875]

LANDMARKS, MEDICAL AND SURGICAL,
By LUTHER HOLDEN, F.R.C.S., Member of the Court of Examiners of
the Royal College of Surgeons. Second Edition, 8vo, 3s. 6d.   [1877]
BY THE SAME AUTHOR.

HUMAN OSTEOLOGY:
Comprising a Description of the Bones, with Delineations of the
Attachments of the Muscles, the General and Microscopical Structure
of Bone, and its Development. Fifth Edition, with 61 Lithographic
Plates and 89 Engravings. 8vo, 16s.   [1878]

PATHOLOGICAL ANATOMY:
Lectures by SAMUEL WILKS, M.D., F.R.S., Physician to, and Lec-
turer on Medicine at, Guy's Hospital; and WALTER MOXON, M.D.,
F.R.C.P., Physician to, and Lecturer on Materia Medica at, Guy's
Hospital. Second Edition, 8vo, with Plates, 18s.   [1875]

PATHOLOGICAL ANATOMY:
A Manual by C. HANDFIELD JONES, M.B., F.R.S., Physician to St.
Mary's Hospital, and EDWARD H. SIEVEKING, M.D., F.R.C.P.,
Physician to St. Mary's Hospital. Edited by J. F. PAYNE, M.D.,
F.R.C.P., Assistant Physician and Lecturer on General Pathology
at St. Thomas's Hospital. Second Edition, crown 8vo, with 195
Engravings, 16s.   [1875]

POST-MORTEM EXAMINATIONS:
a Description and Explanation of the Method of Performing them,
with especial Reference to Medico-Legal Practice. By Professor
RUDOLPH VIRCHOW, of Berlin. Fcap 8vo, 2s. 6d.   [1876]

STUDENT'S GUIDE TO SURGICAL ANATOMY:
a Text-book for the Pass Examination, by E. BELLAMY, F.R.C.S.,
Surgeon and Lecturer on Anatomy at Charing Cross Hospital. Fcap
8vo, with 50 Engravings, 6s. 6d.   [1873]

DIAGRAMS OF THE NERVES OF THE HUMAN BODY,
Exhibiting their Origin, Divisions, and Connexions, with their Distri-
bution, by WILLIAM H. FLOWER, F.R.S., Conservator of Museum,
Royal College of Surgeons. Second Edition, roy. 4to, 12s.   [1872]

MEDICAL ANATOMY,
by FRANCIS SIBSON, M.D., F.R.C.P., F.R.S. Imp. folio, with 21
coloured Plates, cloth, 42s., half-morocco, 50s.   [1869]

PRACTICAL ANATOMY:
a Manual of Dissections by CHRISTOPHER HEATH, F.R.C.S., Surgeon
to University College Hospital, and Holme Professor of Surgery in
University College. Fourth Edition, crown 8vo, with 16 Coloured
Plates and 264 Engravings, 14s.   [1877]

AN ATLAS OF HUMAN ANATOMY:
illustrating most of the ordinary Dissections, and many not usually
practised by the Student. To be completed in 12 or 13 Bi-monthly
Parts, each containing 4 Coloured Plates, with Explanatory Text. By
RICKMAN J. GODLEE, M.S., F.R.C.S., Assistant Surgeon to University
College Hospital, and Senior Demonstrator of Anatomy in University
College. Imp. 4to, 7s. 6d. each Part.   [1877-8]

THE ANATOMIST'S VADE-MECUM:
a System of Human Anatomy by ERASMUS WILSON, F.R.C.S., F.R.S.
Ninth Edition, by G. BUCHANAN, M.A., M.D., Professor of Clinical
Surgery in the University of Glasgow, and HENRY E. CLARK, F.F.P.S.,
Lecturer on Anatomy at the Glasgow Royal Infirmary School of
Medicine. Crown 8vo, with 371 Engravings, 14s.   [1878]

ATLAS OF TOPOGRAPHICAL ANATOMY,
after Plane Sections of Frozen Bodies. By WILHELM BRAUNE,
Professor of Anatomy in the University of Leipzig. Translated by
EDWARD BELLAMY, F.R.C.S., Surgeon to, and Lecturer on Anatomy,
&c., at, Charing Cross Hospital. With 34 Photo-lithographic Plates
and 46 Woodcuts. Large Imp. 8vo, 40s.   [1877]

THE ANATOMICAL REMEMBRANCER;
or, Complete Pocket Anatomist. Eighth Edition, 32mo, 3s. 6d.   [1876]

THE STUDENT'S GUIDE TO THE PRACTICE OF MEDICINE,
by MATTHEW CHARTERIS, M.D., Professor of Medicine in Anderson's
College, and Lecturer on Clinical Medicine in the Royal Infirmary,
Glasgow. With Engravings on Copper and Wood, fcap. 8vo, 6s. 6d. [1877]

THE MICROSCOPE IN MEDICINE,
by LIONEL S. BEALE, M.B., F.R.S., Physician to King's College
Hospital. Fourth Edition, with 86 Plates, 8vo, 21s.   [1877]

HOOPER'S PHYSICIAN'S VADE-MECUM;
or, Manual of the Principles and Practice of Physic, Ninth Edition
by W. A. GUY, M.B., F.R.S., and JOHN HARLEY, M.D., F.R.C.P.
Fcap 8vo, with Engravings, 12s. 6d.   [1874]

A NEW SYSTEM OF MEDICINE;
entitled Recognisant Medicine, or the State of the Sick, by
BHOLANOTH BOSE, M.D., Indian Medical Service. 8vo, 10s. 6d.   [1877]
                        BY THE SAME AUTHOR.
PRINCIPLES OF RATIONAL THERAPEUTICS.
Commenced as an Inquiry into the Relative Value of Quinine and
Arsenic in Ague. 8vo. 4s.   [1877]

THE STUDENT'S GUIDE TO MEDICAL DIAGNOSIS,
by Samuel Fenwick. M.D., F.R.C.P., Physician to the London
Hospital. Fourth Edition, fcap. 8vo, with 106 Engravings, 6s. 6d. [1876]

A MANUAL OF MEDICAL DIAGNOSIS,
by A. W. Barclay, M.D., F.R.C.P., Physician to, and Lecturer on
Medicine at, St. George's Hospital. Third Edition, fcap 8vo, 10s. 6d.
[1876]

CLINICAL MEDICINE:
Lectures and Essays by Balthazar Foster, M.D., F.R.C.P. Lond.,
Professor of Medicine in Queen's College, Birmingham. 8vo, 10s. 6d.
[1874]

CLINICAL STUDIES:
Illustrated by Cases observed in Hospital and Private Practice, by Sir
J. Rose Cormack, M.D., F.R.S.E., Physician to the Hertford British
Hospital of Paris. 2 vols., post 8vo, 20s. [1876]

CLINICAL REMINISCENCES:
By Peyton Blakiston, M.D., F.R.S. Post 8vo, 3s. 6d. [1878]

ROYLE'S MANUAL OF MATERIA MEDICA AND THERAPEUTICS.
Sixth Edition by John Harley, M.D., F.R.C.P., Assistant Physician
to, and Joint Lecturer on Physiology at, St. Thomas's Hospital. Crown
8vo, with 139 Engravings, 15s. [1876]

PRACTICAL THERAPEUTICS:
A Manual by E. J. Waring, M.D., F.R.C.P. Lond. Third Edition,
fcap 8vo. 12s. 6d. [1871]

THE ELEMENTS OF THERAPEUTICS.
A Clinical Guide to the Action of Drugs, by C. Binz, M.D., Professor
of Pharmacology in the University of Bonn. Translated and Edited
with Additions, in Conformity with the British and American Phar-
macopœias, by Edward I. Sparks, M.A., M.B. Oxon., formerly
Radcliffe Travelling Fellow. Crown 8vo, 8s. 6d. [1877]

THE STUDENT'S GUIDE TO MATERIA MEDICA,
by John C. Thorowgood, M.D., F.R.C.P. Lond., Physician to the
City of London Hospital for Diseases of the Chest. Fcap 8vo, with
Engravings, 6s. 6d. [1874]

MATERIA MEDICA AND THERAPEUTICS:
(Vegetable Kingdom), by Charles D. F. Phillips, M.D., F.R.C.S.E.
8vo, 15s. [1876]

DENTAL MATERIA MEDICA AND THERAPEUTICS,
Elements of, by James Stocken, L.D.S.R.C.S., Lecturer on Dental
Materia Medica and Therapeutics to the National Dental Hospital.
Second Edition, Fcap 8vo, 6s. 6d. [1878]

THE DISEASES OF CHILDREN:
A Practical Manual, with a Formulary, by Edward Ellis, M.D.,
late Senior Physician to the Victoria Hospital for Children. Third
Edition, crown 8vo, 7s. 6d. [1878]

THE WASTING DISEASES OF CHILDREN,
by EUSTACE SMITH, M.D., F.R.C.P. Lond., Physician to the King of
the Belgians, Physician to the East London Hospital for Children.
Third Edition, post 8vo. [In the Press.]

BY THE SAME AUTHOR,
CLINICAL STUDIES OF DISEASE IN CHILDREN.
Post 8vo, 7s. 6d. [1876]

INFANT FEEDING AND ITS INFLUENCE ON LIFE;
or, the Causes and Prevention of Infant Mortality, by CHARLES H. F.
ROUTH, M.D., Senior Physician to the Samaritan Hospital for Women
and Children. Third Edition, fcap 8vo, 7s. 6d. [1876]

COMPENDIUM OF CHILDREN'S DISEASES:
A Handbook for Practitioners and Students, by JOHANN STEINER,
M.D., Professor in the University of Prague. Translated from the
Second German Edition by LAWSON TAIT, F.R.C.S., Surgeon to the
Birmingham Hospital for Women. 8vo, 12s. 6d. [1874]

THE DISEASES OF CHILDREN:
Essays by WILLIAM HENRY DAY, M.D., Physician to the Samaritan
Hospital for Diseases of Women and Children. Second Edition, fcap 8vo.
[In the Press.]

PUERPERAL DISEASES:
Clinical Lectures by FORDYCE BARKER, M.D., Obstetric Physician
to Bellevue Hospital, New York. 8vo, 15s. [1874]

THE STUDENT'S GUIDE TO THE PRACTICE OF MIDWIFERY,
by D. LLOYD ROBERTS, M.D., F.R.C.P., Physician to St. Mary's Hospital, Manchester. Second Edition, fcap. 8vo, with 95 Engravings.
[In the Press.]

OBSTETRIC MEDICINE AND SURGERY,
Their Principles and Practice, by F. H. RAMSBOTHAM, M.D., F.R.C.P.
Fifth Edition, 8vo, with 120 Plates, 22s. [1867]

OBSTETRIC SURGERY:
A Complete Handbook, giving Short Rules of Practice in every Emer-
gency, from the Simplest to the most Formidable Operations connected
with the Science of Obstetricy, by CHARLES CLAY, Ext.L.R.C.P. Lond.,
L.R.C.S.E., late Senior Surgeon and Lecturer on Midwifery, St.
Mary's Hospital, Manchester. Fcap 8vo, with 91 Engravings, 6s. 6d.
[1874]

SCHROEDER'S MANUAL OF MIDWIFERY,
including the Pathology of Pregnancy and the Puerperal State.
Translated by CHARLES H. CARTER, B.A., M.D. 8vo, with Engravings, 12s. 6d. [1873]

A HANDBOOK OF UTERINE THERAPEUTICS,
and of Diseases of Women, by E. J. TILT, M.D., M.R.C.P. Fourth
Edition, post 8vo, 10s. [1878]

BY THE SAME AUTHOR,
THE CHANGE OF LIFE
in Health and Disease: a Practical Treatise on the Nervous and other
Affections incidental to Women at the Decline of Life. Third Edition,
8vo, 10s. 6d. [1870]

OBSTETRIC OPERATIONS,
including the Treatment of Hæmorrhage, and forming a Guide to the
Management of Difficult Labour; Lectures by ROBERT BARNES, M.D.,
F.R.C.P., Obstetric Physician and Lecturer on Obstetrics and the Diseases of Women and Children at St. George's Hospital. Third Edition,
8vo, with 124 Engravings, 18s.   [1875]

BY THE SAME AUTHOR,
MEDICAL AND SURGICAL DISEASES OF WOMEN :
a Clinical History. Second Edition, 8vo, with 181 Engravings, 28s.
                                                                 [1878]
OBSTETRIC APHORISMS:
for the Use of Students commencing Midwifery Practice by J. G.
SWAYNE, M.D., Consulting Physician-Accoucheur to the Bristol
General Hospital, and Lecturer on Obstetric Medicine at the Bristol
Medical School. Sixth Edition, fcap 8vo, with Engravings, 3s. 6d. [1876]

DISEASES OF THE OVARIES :
their Diagnosis and Treatment, by T. SPENCER WELLS, F.R.C.S.,
Surgeon to the Queen's Household and to the Samaritan Hospital.
8vo, with about 150 Engravings, 21s.   [1872]

PRACTICAL GYNÆCOLOGY :
A Handbook of the Diseases of Women, by HEYWOOD SMITH, M.D.
Oxon., Physician to the Hospital for Women and to the British Lying-
in Hospital. With Engravings, crown 8vo, 5s. 6d.   [1877]

RUPTURE OF THE FEMALE PERINEUM,
Its treatment, immediate and remote, by GEORGE G. BANTOCK, M.D.,
Surgeon (for In-patients) to the Samaritan Free Hospital for Women
and Children. With 2 plates, 8vo, 3s. 6d.   [1878.]

INFLUENCE OF POSTURE ON WOMEN
In Gynecic and Obstetric Practice, by J. H. AVELING, M.D., Physi-
cian to the Chelsea Hospital for Women, Vice-President of the
Obstetrical Society of London. 8vo, 6s.   [1878]

HANDBOOK FOR NURSES FOR THE SICK,
by ZEPHERINA P. VEITCH. Second Edition, crown 8vo, 3s. 6d.   [1876]

A MANUAL FOR HOSPITAL NURSES
and others engaged in Attending on the Sick by EDWARD J. DOM-
VILLE, L.R.C.P., M.R.C.S., Surgeon to the Exeter Lying-in Charity.
Third Edition, crown 8vo, 2s. 6d.   [1878]

THE NURSE'S COMPANION :
A Manual of General and Monthly Nursing, by CHARLES J. CULLING-
WORTH, Surgeon to St. Mary's Hospital, Manchester. Fcap. 8vo,
2s. 6d.   [1876]

LECTURES ON NURSING,
by WILLIAM ROBERT SMITH, M.B., Honorary Medical Officer,
Hospital for Sick Children, Sheffield. Second Edition, with 26 Engravings. Post 8vo, 6s.   [1878]

A COMPENDIUM OF DOMESTIC MEDICINE

and Companion to the Medicine Chest; intended as a Source of Easy Reference for Clergymen, and for Families residing at a Distance from Professional Assistance, by JOHN SAVORY, M.S.A. Ninth Edition, 12mo, 5s. [1878]

HOSPITAL MORTALITY

being a Statistical Investigation of the Returns of the Hospitals of Great Britain and Ireland for fifteen years, by LAWSON TAIT, F.R.C.S., F.S.S. 8vo, 8s. 6d. [1877]

THE COTTAGE HOSPITAL:

Its Origin, Progress, Management, and Work, by HENRY C. BURDETT, the Seaman's Hospital, Greenwich. With Engravings, crown 8vo, 7s. 6d. [1877]

WINTER COUGH:

(Catarrh, Bronchitis, Emphysema, Asthma), Lectures by HORACE DOBELL, M.D., Consulting Physician to the Royal Hospital for Diseases of the Chest. Third Edition, with Coloured Plates, 8vo, 1s. 6d. [1875]

DISEASES OF THE CHEST:

Contributions to their Clinical History, Pathology, and Treatment, by A. T. H. WATERS, M.D., F.R.C.P., Physician to the Liverpool Royal Infirmary. Second Edition, 8vo, with Plates, 15s. [1873]

CONSUMPTION:

Its Nature, Symptoms, Causes, Prevention, Curability, and Treatment. By PETER GOWAN, M.D., B. Sc., late Physician and Surgeon in Ordinary to the King of Siam. Crown 8vo. 5s. [1878]

NOTES ON ASTHMA;

its Forms and Treatment, by JOHN C. THOROWGOOD, M.D. Lond., F.R.C.P., Physician to the Hospital for Diseases of the Chest, Victoria Park. Third Edition, crown 8vo, 4s. 6d. [1878]

ASTHMA

Its Pathology and Treatment, by J. B. BERKART, M.D., Assistant Physician to the City of London Hospital for Diseases of the Chest. 8vo, 7s. 6d. [1878]

PROGNOSIS IN CASES OF VALVULAR DISEASE OF THE

Heart, by THOMAS B. PEACOCK, M.D., F.R.C.P., Honorary Consulting Physician to St. Thomas's Hospital. 8vo, 3s. 6d. [1877]

DISEASES OF THE HEART:

Their Pathology, Diagnosis, Prognosis, and Treatment (a Manual), by ROBERT H. SEMPLE, M.D., F.R.C.P., Physician to the Hospital for Diseases of the Throat. 8vo, 8s. 6d. [1875]

CHRONIC DISEASE OF THE HEART:

Its Bearings upon Pregnancy, Parturition and Childbed. By ANGUS MACDONALD, M.D., F.R.S.E., Physician to, and Clinical Lecturer on the Diseases of Women at, the Edinburgh Royal Infirmary. With Engravings, 8vo. [1878]

PHTHISIS:

In a series of Clinical Studies, by AUSTIN FLINT, M.D., Professor of the Principles and Practice of Medicine and of Clinical Medicine in the Bellevue Hospital Medical College. 8vo, 16s.   [1875]

BY THE SAME AUTHOR,

A MANUAL OF PERCUSSION AND AUSCULTATION,

of the Physical Diagnosis of Diseases of the Lungs and Heart, and of Thoracic Aneurism. Post 8vo, 6s. 6d.   [1876]

GROWTHS IN THE LARYNX,

with Reports and an Analysis of 100 consecutive Cases treated since the Invention of the Laryngoscope by MORELL MACKENZIE, M.D. Lond., M.R.C.P., Physician to the Hospital for Diseases of the Throat. 8vo, with Coloured Plates, 12s. 6d.   [1871]

DISEASES OF THE HEART AND AORTA,

By THOMAS HAYDEN, F.K.Q.C.P. Irel., Physician to the Mater Misericordiæ Hospital, Dublin. With 80 Engravings. 8vo, 25s.   [1875]

DISEASES OF THE HEART

and of the Lungs in Connexion therewith—Notes and Observations by THOMAS SHAPTER, M.D., F.R.C.P. Lond., Senior Physician to the Devon and Exeter Hospital. 8vo, 7s. 6d.   [1874]

DISEASES OF THE HEART AND AORTA:

Clinical Lectures by GEORGE W. BALFOUR, M.D., F.R.C.P., Physician to, and Lecturer on Clinical Medicine in, the Royal Infirmary, Edinburgh. 8vo, with Engravings. 12s. 6d.   [1876]

PHYSICAL DIAGNOSIS OF DISEASES OF THE HEART.

Lectures by ARTHUR E. SANSOM, M.D., F.R.C.P., Assistant Physician to the London Hospital. Second Edition, with Engravings, fcap. 8vo, 4s. 6d.   [1876]

TRACHEOTOMY,

especially in Relation to Diseases of the Larynx and Trachea, by PUGIN THORNTON, M.R.C.S., late Surgeon to the Hospital for Diseases of the Throat. With Photographic Plates and Woodcuts, 8vo, 5s. 6d.   [1876]

SORE THROAT:

Its Nature, Varieties, and Treatment, including the Connexion between Affections of the Throat and other Diseases. By PROSSER JAMES, M.D., Lecturer on Materia Medica and Therapeutics at the London Hospital, Physician to the Hospital for Diseases of the Throat. Third Edition, with Coloured Plates, 5s. 6d.   [1878]

WINTER AND SPRING

on the Shores of the Mediterranean. By HENRY BENNET, M.D. Fifth Edition, post 8vo, with numerous Plates, Maps, and Engravings, 12s. 6d.   [1871]

BY THE SAME AUTHOR,

TREATMENT OF PULMONARY CONSUMPTION

by Hygiene, Climate, and Medicine. Second Edition, 8vo, 5s.   [1871]

PRINCIPAL HEALTH RESORTS
of Europe and Africa, and their Use in the Treatment of Chronic
Diseases. A Handbook by THOMAS MORE MADDEN, M.D., M.R.I.A.,
Vice-President of the Dublin Obstetrical Society. 8vo, 10s. [1876]

THE BATH THERMAL WATERS:
Historical, Social, and Medical, by JOHN KENT SPENDER, M.D.,
Surgeon to the Mineral Water Hospital, Bath. With an Appendix
on the Climate of Bath by the Rev. L. BLOMEFIELD, M.A., F.L.S.,
F.G.S. 8vo, 7s. 6d. [1877]

ENDEMIC DISEASES OF TROPICAL CLIMATES,
with their Treatment, by JOHN SULLIVAN, M.D., M.R.C.P. Post 8vo,
6s. [1877]

DISEASES OF TROPICAL CLIMATES
and their Treatment: with Hints for the Preservation of Health in the
Tropics, by JAMES A. HORTON, M.D., Surgeon-Major, Army Medical
Department. Post 8vo, 12s. 6d. [1874]

HEALTH IN INDIA FOR BRITISH WOMEN
and on the Prevention of Disease in Tropical Climates by EDWARD J.
TILT, M.D., Consulting Physician-Accoucheur to the Farringdon
General Dispensary. Fourth Edition, crown 8vo, 5s. [1875]

BURDWAN FEVER,
or the Epidemic Fever of Lower Bengal (Causes, Symptoms, and
Treatment), by GOPAUL CHUNDER ROY, M.D., Surgeon Bengal
Establishment. New Edition, 8vo, 5s. [1876]

BAZAAR MEDICINES OF INDIA
and Common Medical Plants: Remarks on their Uses, with Full Index
of Diseases, indicating their Treatment by these and other Agents pro-
curable throughout India, &c., by EDWARD J. WARING, M.D., F.R.C.P.
Lond., Retired Surgeon H.M. Indian Army. Third Edition. Fcap
8vo, 5s. [1875]

SOME AFFECTIONS OF THE LIVER
and Intestinal Canal; with Remarks on Ague and its Sequelæ, Scurvy,
Purpura, &c., by STEPHEN H. WARD, M.D. Lond., F.R.C.P., Physician
to the Seamen's Hospital, Greenwich. 8vo, 7s. [1872]

DISEASES OF THE LIVER:
Lettsomian Lectures for 1872 by S. O. HABERSHON, M.D., F.R.C.P.,
Senior Physician to Guy's Hospital. Post 8vo, 3s. 6d. [1872]

BY THE SAME AUTHOR,
DISEASES OF THE STOMACH: DYSPEPSIA.
Second Edition, crown 8vo, 5s.

BY THE SAME AUTHOR,
PATHOLOGY OF THE PNEUMOGASTRIC NERVE,
being the Lumleian Lectures for 1876. Post 8vo, 3s. 6d. [1877]

BY THE SAME AUTHOR,
DISEASES OF THE ABDOMEN,
comprising those of the Stomach and other parts of the Alimentary
Canal, Œsophagus, Cæcum, Intestines, and Peritoneum. Third
Edition, with 5 Plates, 8vo, 21s. [1878]

FUNCTIONAL NERVOUS DISORDERS:
Studies by C. HANDFIELD JONES, M.B., F.R.C.P., F.R.S., Physician
to St. Mary's Hospital. Second Edition, 8vo, 18s. [1870]

LECTURES ON DISEASES OF THE NERVOUS SYSTEM,
by SAMUEL WILKS, M.D., F.R.S., Physician to, and Lecturer on
Medicine at, Guy's Hospital. 8vo, 15s. [1878]

NERVOUS DISEASES:
their Description and Treatment, by ALLEN McLANE HAMILTON, M.D.,
Physician at the Epileptic and Paralytic Hospital, Blackwell's Island,
New York City. Roy. 8vo, with 53 Illustrations, 14s. [1878]

NUTRITION IN HEALTH AND DISEASE:
A Contribution to Hygiene and to Clinical Medicine. By HENRY
BENNET, M.D. Third Edition. 8vo, 7s. Cheap Edition, Fcap. 8vo,
2s. 6d. [1877]

FOOD AND DIETETICS,
Physiologically and Therapeutically Considered. By FREDERICK W.
PAVY, M.D., F.R.S., Physician to Guy's Hospital. Second Edition,
8vo, 15s. [1875]

HEADACHES:
their Causes, Nature, and Treatment. By WILLIAM H. DAY, M.D.,
Physician to the Samaritan Free Hospital for Women and Children.
Second Edition, crown 8vo, with Engravings. 6s. 6d. [1878]

IMPERFECT DIGESTION:
its Causes and Treatment by ARTHUR LEARED, M.D., F.R.C.P.,
Senior Physician to the Great Northern Hospital. Sixth Edition,
fcap 8vo, 4s. 6d. [1875]

MEGRIM, SICK-HEADACHE,
and some Allied Disorders: a Contribution to the Pathology of Nerve-
Storms, by EDWARD LIVEING, M.D. Cantab., F.R.C.P., Hon. Fellow
of King's College, London. 8vo, with Coloured Plate, 15s. [1873]

NEURALGIA AND KINDRED DISEASES
of the Nervous System: their Nature, Causes, and Treatment, with a
series of Cases, by JOHN CHAPMAN, M.D., M.R.C.P. 8vo, 14s. [1873]

THE SYMPATHETIC SYSTEM OF NERVES,
and their Functions as a Physiological Basis for a Rational System of
Therapeutics by EDWARD MERYON, M.D., F.R.C.P., Physician to the
Hospital for Diseases of the Nervous System. 8vo, 3s. 6d. [1872]

RHEUMATIC GOUT,
or Chronic Rheumatic Arthritis of all the Joints; a Treatise by
ROBERT ADAMS, M.D., M.R.I.A., late Surgeon to H.M. the Queen in
Ireland, and Regius Professor of Surgery in the University of Dublin.
Second Edition, 8vo, with Atlas of Plates, 21s. [1872]

GOUT, RHEUMATISM,
and the Allied Affections; a Treatise by PETER HOOD, M.D. Crown
8vo, 10s. 6d. [1871]

RHEUMATISM:

Notes by JULIUS POLLOCK, M.D., F.R.C.P., Senior Physician to, and Lecturer on Medicine at, Charing Cross Hospital. Fcap. 8vo, 2s. 6d.
[1878.]

CANCER:

its varieties, their Histology and Diagnosis, by HENRY ARNOTT, F.R.C.S., late Assistant-Surgeon to, and Lecturer on Morbid Anatomy at, St. Thomas's Hospital. 8vo, with 5 Plates and 22 Engravings, 5s. 6d. [1872]

CANCEROUS AND OTHER INTRA-THORACIC GROWTHS:

their Natural History and Diagnosis, by J. RISDON BENNETT, M.D., F.R.C.P., Member of the General Medical Council. Post 8vo, with Plates, 8s. [1872]

CERTAIN FORMS OF CANCER,

with a New and successful Mode of Treating it, to which is prefixed a Practical and Systematic Description of all the varieties of this Disease, by ALEX. MARSDEN, M.D., F.R.C.S.E., Consulting Surgeon to the Royal Free Hospital, and Senior Surgeon to the Cancer Hospital. Second Edition, with Coloured Plates, 8vo, 8s. 6d. [1873]

ATLAS OF SKIN DISEASES:

a series of Illustrations, with Descriptive Text and Notes upon Treatment. By TILBURY FOX, M.D., F.R.C.P., Physician to the Department for Skin Diseases in University College Hospital. With 72 Coloured Plates, royal 4to, half morocco, £6 6s. [1877]

DISEASES OF THE SKIN:

a System of Cutaneous Medicine by ERASMUS WILSON, F.R.C.S., F.R.S. Sixth Edition, 8vo, 18s., with Coloured Plates, 36s. [1867]

BY THE SAME AUTHOR,

LECTURES ON EKZEMA

and Ekzematous Affections: with an Introduction on the General Pathology of the Skin, and an Appendix of Essays and Cases. 8vo, 10s. 6d. [1870]

ALSO,

LECTURES ON DERMATOLOGY:

delivered at the Royal College of Surgeons, 1870, 6s. ; 1871-3, 10s. 6d., 1874-5, 10s. 6d. ; 1876-8, 10s. 6d.

ECZEMA:

by McCALL ANDERSON, M.D., Professor of Clinical Medicine in the University of Glasgow. Third Edition, 8vo, with Engravings, 7s. 6d. [1874]

BY THE SAME AUTHOR,

PARASITIC AFFECTIONS OF THE SKIN

Second Edition, 8vo, with Engravings, 7s. 6d. [1868]

PSORIASIS OR LEPRA,

by GEORGE GASKOIN, M.R.C.S., Surgeon to the British Hospital for Diseases of the Skin. 8vo, 5s. [1875]

MYCETOMA;
>   or, the Fungus Disease of India, by H. VANDYKE CARTER, M.D., Sur-
>   geon-Major H.M. Indian Army. 4to, with 11 Coloured Plates, 42s.
>   [1874]

CERTAIN ENDEMIC SKIN AND OTHER DISEASES
>   of India and Hot Climates generally, by TILBURY FOX, M.D., F.R.C.P.,
>   and T. FARQUHAR, M.D. (Published under the sanction of the Secre-
>   tary of State for India in Council). 8vo, 10s. 6d.   [1876]

DISEASES OF THE SKIN,
>   in Twenty-four Letters on the Principles and Practice of Cutaneous
>   Medicine, by HENRY EVANS CAUTY, M.R.C.S., Surgeon to the Liver-
>   pool Dispensary for Diseases of the Skin, 8vo, 12s. 6d.   [1874]

THE HAIR IN HEALTH AND DISEASE,
>   by E. WYNDHAM COTTLE, F.R.C.S., Senior Assistant Surgeon to the
>   Hospital for Diseases of the Skin, Blackfriars. Fcap. 8vo, 2s. 6d.   [1877]

WORMS:
>   a Series of Lectures delivered at the Middlesex Hospital on Practical
>   Helminthology by T. SPENCER COBBOLD, M.D., F.R.S. Post 8vo,
>   5s.   [1872]

THE LAWS AFFECTING MEDICAL MEN:
>   a Manual by ROBERT G. GLENN, LL.B., Barrister-at-Law; with a
>   Chapter on Medical Etiquette by Dr. A. CARPENTER. 8vo, 14s.   [1871]

MEDICAL JURISPRUDENCE,
>   Its Principles and Practice, by ALFRED S. TAYLOR, M.D., F.R.C.P.,
>   F.R.S. Second Edition, 2 vols., 8vo, with 189 Engravings, £1 11s. 6d.
>   [1873]

BY THE SAME AUTHOR,

A MANUAL OF MEDICAL JURISPRUDENCE.
>   Ninth Edition. Crown 8vo, with Engravings, 14s.   [1874]

ALSO,

POISONS,
>   in Relation to Medical Jurisprudence and Medicine. Third Edition,
>   crown 8vo, with 104 Engravings, 16s.   [1875]

MEDICAL JURISPRUDENCE:
>   Lectures by FRANCIS OGSTON, M.D., Professor of Medical Juris-
>   prudence and Medical Logic in the University of Aberdeen. Edited
>   by FRANCIS OGSTON, Jun., M.D., Assistant to the Professor of
>   Medical Jurisprudence and Lecturer on Practical Toxicology in the
>   University of Aberdeen. 8vo, with 12 Copper Plates, 18s.   [1878]

A TOXICOLOGICAL CHART,
>   exhibiting at one View the Symptoms, Treatment, and mode of
>   Detecting the various Poisons—Mineral, Vegetable, and Animal:
>   with Concise Directions for the Treatment of Suspended Animation,
>   by WILLIAM STOWE, M.R.C.S.E. Thirteenth Edition, 2s.; on
>   roller, 5s.   [1872]

A HANDY-BOOK OF FORENSIC MEDICINE AND TOXICOLOGY,
by W. BATHURST WOODMAN, M.D., F.R.C.P., Assistant Physician
and Co-Lecturer on Physiology and Histology at the London Hospital;
and C. MEYMOTT TIDY, M.D., F.C.S.. Professor of Chemistry and of
Medical Jurisprudence and Public Health at the London Hospital.
With 8 Lithographic Plates and 116 Engravings, 8vo, 31s. 6d. [1877]

THE MEDICAL ADVISER IN LIFE ASSURANCE,
by EDWARD HENRY SIEVEKING, M.D., F.R.C.P., Physician to St.
Mary's and the Lock Hospitals; Physician-Extraordinary to the
Queen; Physician-in-Ordinary to the Prince of Wales, &c. Crown
8vo. 6s. [1874]

IDIOCY AND IMBECILITY,
by WILLIAM W. IRELAND, M.D., Medical Superintendent of the
Scottish National Institution for the Education of Imbecile Children
at Larbert, Stirlingshire. With Engravings, 8vo, 14s. [1877]

PSYCHOLOGICAL MEDICINE:
a Manual, containing the Lunacy Laws, the Nosology, Ætiology,
Statistics, Description, Diagnosis, Pathology (including Morbid Histology), and Treatment of Insanity, by J. C. BUCKNILL, M.D.,
F.R.S., and D. H. TUKE, M.D., F.R.C.P. Third Edition, 8vo, with
10 Plates and 34 Engravings, 25s. [1873]

MADNESS:
in its Medical, Legal, and Social Aspects, Lectures by EDGAR
SHEPPARD, M.D., M.R.C.P., Professor of Psychological Medicine in
King's College; one of the Medical Superintendents of the Colney
Hatch Lunatic Asylum. 8vo, 6s. 6d. [1873]

HANDBOOK OF LAW AND LUNACY;
or, the Medical Practitioner's Complete Guide in all Matters relating
to Lunacy Practice, by J. T. SABBEN, M.D., and J. H. BALFOUR
BROWNE, Barrister-at-Law. 8vo, 5s. [1872]

INFLUENCE OF THE MIND UPON THE BODY
in Health and Disease, Illustrations designed to elucidate the Action
of the Imagination, by DANIEL HACK TUKE, M.D., F.R.C.P.
8vo, 14s. [1872]

A MANUAL OF PRACTICAL HYGIENE,
by E. A. PARKES, M.D., F.R.S. Fifth Edition, by F. DE CHAUMONT.
M.D., Professor of Military Hygiene in the Army Medical School.
8vo, with 9 Plates and 112 Engravings, 18s. [1875]

A HANDBOOK OF HYGIENE AND SANITARY SCIENCE,
by GEORGE WILSON, M.A., M.D., Medical Officer of Health for Mid-
Warwickshire. Third Edition, post 8vo, with Engravings, 10s. 6d. [1877]

MICROSCOPICAL EXAMINATION OF DRINKING WATER:
A Guide, by JOHN D. MACDONALD, M.D., F.R.S., Assistant Professor of Naval Hygiene, Army Medical School. 8vo, with 24 Plates,
7s. 6d. [1875]

HANDBOOK OF MEDICAL AND SURGICAL ELECTRICITY,
> by HERBERT TIBBITS, M.D., F.R.C.P.E.. Medical Superintendent of the National Hospital for the Paralysed and Epileptic. Second Edition 8vo, with 95 Engravings, 9s.   [1877]

BY THE SAME AUTHOR.
A MAP OF ZIEMSSEN'S MOTOR POINTS OF THE HUMAN BODY:
> a Guide to Localised Electrisation. Mounted on Rollers, 35 × 21. With 20 Illustrations. 5s.   [1877]

CLINICAL USES OF ELECTRICITY;
> Lectures delivered at University College Hospital by J. RUSSELL REYNOLDS, M.D. Lond., F.R.C.P., F.R.S., Professor of Medicine in University College. Second Edition, post 8vo, 3s. 6d.   [1873]

MEDICO-ELECTRIC APPARATUS:
> A Practical Description of every Form in Modern Use, with Plain Directions for Mounting, Charging, and Working, by SALT & SON, Birmingham. Second Edition, revised and enlarged, with 33 Engravings, 8vo, 2s. 6d.   [1877]

A DICTIONARY OF MEDICAL SCIENCE;
> containing a concise explanation of the various subjects and terms of Medicine, &c.; Notices of Climate and Mineral Waters; Formulæ for Officinal, Empirical, and Dietetic Preparations; with the Accentuation and Etymology of the terms and the French and other Synonyms, by ROBLEY DUNGLISON. M.D.. LL.D. New Edition, royal 8vo, 28s.   [1874]

A MEDICAL VOCABULARY;
> being an Explanation of all Terms and Phrases used in the various Departments of Medical Science and Practice. giving their derivation, meaning, application, and pronunciation, by ROBERT G. MAYNE, M.D., LL.D. Fourth Edition. fcap 8vo, 10s.   [1875]

ATLAS OF OPHTHALMOSCOPY,
> by R. LIEBREICH, Ophthalmic Surgeon to St. Thomas's Hospital. Translated into English by H. ROSBOROUGH SWANZY, M.B. Dub Second Edition, containing 59 Figures, 4to, £1 10s.   [1870]

DISEASES OF THE EYE:
> a Manual by C. MACNAMARA, F.R.C.S., Surgeon to Westminster Hospital. Third Edition, fcap. 8vo, with Coloured Plates and Engravings, 12s. 6d.   [1876]

DISEASES OF THE EYE:
> A Practical Treatise by HAYNES WALTON, F.R.C.S., Surgeon to St. Mary's Hospital and in charge of its Ophthalmological Department. Third Edition, 8vo, with 3 Plates and nearly 300 Engravings, 25s.   [1875]

HINTS ON OPHTHALMIC OUT-PATIENT PRACTICE,
> by CHARLES HIGGENS, F.R.C.S.. Ophthalmic Assistant Surgeon to, and Lecturer on Ophthalmology at, Guy's Hospital. 87 pp., fcap. 8vo, 2s. 6d.   [1877]

OPHTHALMIC MEDICINE AND SURGERY:
a Manual by T. WHARTON JONES, F.R.C.S., F.R.S., Professor of Ophthalmic Medicine and Surgery in University College. Third Edition, fcap. 8vo, with 9 Coloured Plates and 173 Engravings, 12s. 6d.   [1865]

DISEASES OF THE EYE:
A Treatise by J. SOELBERG WELLS, F.R.C.S., Ophthalmic Surgeon to King's College Hospital and Surgeon to the Royal London Ophthalmic Hospital. Third Edition, 8vo, with Coloured Plates and Engravings, 25s.   [1873]

BY THE SAME AUTHOR,

LONG, SHORT, AND WEAK SIGHT,
and their Treatment by the Scientific use of Spectacles. Fourth Edition, 8vo, 6s.   [1873]

A SYSTEM OF DENTAL SURGERY,
by JOHN TOMES, F.R.S., and CHARLES S. TOMES, M.A., F.R.S., Lecturer on Dental Anatomy and Physiology at the Dental Hospital of London. Second Edition, fcap 8vo, with 268 Engravings, 14s.   [1873]

DENTAL ANATOMY, HUMAN AND COMPARATIVE:
A Manual, by CHARLES S. TOMES, M.A., F.R.S., Lecturer on Dental Anatomy and Physiology at the Dental Hospital of London. With 179 Engravings, crown 8vo, 10s. 6d.   [1876]

A MANUAL OF DENTAL MECHANICS,
with an Account of the Materials and Appliances used in Mechanical Dentistry, by OAKLEY COLES, L.D.S., R.C.S., Surgeon-Dentist to the Hospital for Diseases of the Throat. Second Edition, crown 8vo, with 140 Engravings, 7s. 6d.   [1876]

HANDBOOK OF DENTAL ANATOMY
and Surgery for the use of Students and Practitioners by JOHN SMITH, M.D., F.R.S. Edin., Surgeon-Dentist to the Queen in Scotland. Second Edition, fcap 8vo, 4s. 6d.   [1871]

STUDENT'S GUIDE TO DENTAL ANATOMY AND SURGERY,
by HENRY SEWILL, M.R.C.S., L.D.S., Dentist to the West London Hospital. With 77 Engravings, fcap. 8vo, 5s. 6d.   [1876]

OPERATIVE DENTISTRY:
A Practical Treatise, by JONATHAN TAFT, D.D.S., Professor of Operative Dentistry in the Ohio College of Dental Surgery. Third Edition, thoroughly revised, with many additions, and 134 Engravings, 8vo, 18s.   [1877]

DENTAL CARIES
and its Causes: an Investigation into the influence of Fungi in the Destruction of the Teeth, by Drs. LEBER and ROTTENSTEIN. Translated by H. CHANDLER, D.M.D., Professor in the Dental School of Harvard University. With Illustrations, royal 8vo, 5s.   [1878.]

EPIDEMIOLOGY;
or, the Remote Cause of Epidemic Diseases in the Animal and in the Vegetable Creation, by JOHN PARKIN, M.D., F.R.C.P.E. Part I, Contagion—Modern Theories—Cholera—Epizootics. 8vo, 5s.   [1873]

The following CATALOGUES issued by Messrs CHURCHILL will be forwarded post free on application:

1. *Messrs Churchill's General List of nearly 600 works on Medicine, Surgery, Midwifery, Materia Medica, Hygiene, Anatomy, Physiology, Chemistry, &c., &c., with a complete Index to their Titles, for easy reference.*

N.B.—*This List includes Nos. 2 and 3.*

2. *Selection from Messrs Churchill's General List, comprising all recent Works published by them on the Art and Science of Medicine.*

3. *A descriptive List of Messrs Churchill's Works on Chemistry, Pharmacy, Botany, Photography, Zoology, and other branches of Science.*

4. *Messrs Churchill's Red-Letter List, giving the Titles of forthcoming New Works and New Editions.*

[Published every October.]

5. *The Medical Intelligencer, an Annual List of New Works and New Editions published by Messrs J. & A. Churchill, together with Particulars of the Periodicals issued from their House.*

[Sent in January of each year to every Medical Practitioner in the United Kingdom whose name and address can be ascertained. A large number are also sent to the United States of America, Continental Europe, India, and the Colonies.]

Messrs CHURCHILL have a special arrangement with MESSRS LINDSAY & BLAKISTON, OF PHILADELPHIA, in accordance with which that Firm act as their Agents for the United States of America, either keeping in Stock most of Messrs CHURCHILL'S Books, or reprinting them on Terms advantageous to Authors. Many of the Works in this Catalogue may therefore be easily obtained in America.

PRINTED BY J. E. ADLARD, BARTHOLOMEW CLOSE.

www.ingramcontent.com/pod-product-compliance
Lightning Source LLC
Chambersburg PA
CBHW021502210326
41599CB00012B/1103